10	11	12	13	14					2 He 4.0026
			BORON 5 B 10.811	CARBON 6 C 12.011	NITROGEN 7 N 14.007	OXYGEN 8 O 15.999	FLUORINE 9 F 18.998	NEON 10 Ne 20.180	
			ALUMINIUM 13 Al 26.982	SILICON 14 Si 28.086	PHOSPHORUS 15 P 30.974	SULFUR 16 S 32.065	CHLORINE 17 Cl 35.453	ARGON 18 Ar 39.948	
NICKEL 28 Ni 58.693	COPPER 29 Cu 63.546	ZINC 30 Zn 65.39	GALLIUM 31 Ga 69.723	GERMANIUM 32 Ge 72.61	ARSENIC 33 As 74.922	SELENIUM 34 Se 78.96	BROMINE 35 Br 79.904	KRYPTON 36 Kr 83.80	
PALLADIUM 46 Pd 106.42	SILVER 47 Ag 107.87	CADMIUM 48 Cd 112.41	INDIUM 49 In 114.82	TIN 50 Sn 118.71	ANTIMONY 51 Sb 121.76	TELLURIUM 52 Te 127.60	IODINE 53 I 126.90	XENON 54 Xe 131.29	
PLATINUM 78 Pt 195.08	GOLD 79 Au 196.97	MERCURY 80 Hg 200.59	THALLIUM 81 Tl 204.38	LEAD 82 Pb 207.2	BISMUTH 83 Bi 208.98	POLONIUM 84 Po [209]	ASTATINE 85 At [210]	RADON 86 Rn [222]	
UNUNNILLIUM 110 Uun [271]	ROENTGENIUM 111 Rg [272]	COPERNICIUM 112 Cn [285]	UNUNTRIUM 113 Uut [286]	UNUNQUADIUM 114 Uuq [289]					

GADOLINIUM 64 Gd 157.25	TERBIUM 65 Tb 158.93	DYSPROSIUM 66 Dy 162.50	HOLMIUM 67 Ho 164.93	ERBIUM 68 Er 167.26	THULIUM 69 Tm 168.93	YTTERBIUM 70 Yb 173.04
CURIUM 96 Cm [247]	BERKELIUM 97 Bk [247]	CALIFORNIUM 98 Cf [251]	EINSTEINIUM 99 Es [252]	FERMIUM 100 Fm [257]	MENDELEVIUM 101 Md [258]	NOBELIUM 102 No [259]

The periodic table

QUADRILLE

The knowledge.
The periodic table | Paul Strathern

Publishing consultant Jane O'Shea
Editor Simon Willis
Creative director Helen Lewis
Art direction & design Claire Peters
Illustrator Claire Peters
Production Vincent Smith, Tom Moore

First published in 2015 by
Quadrille Publishing Limited
www.quadrille.co.uk

Quadrille is an imprint of Hardie Grant.
www.hardiegrant.com.au

Text © 2015 Paul Strathern
Design and layout © 2015
Quadrille Publishing Limited

*Additional table on page 30 © Emilio
Segre Visual Archives/American Institute
of Physics/Science Photo Library*

Cataloguing in Publication Data:
a catalogue record for this book
is available from the British Library.

ISBN 978 184949 626 1
Printed in the UK

1 DISCOVERING THE PERIODIC TABLE 6
2 HOW THE PERIODIC TABLE WORKS 26
3 HOW ATOMS REACT WITH ONE ANOTHER 44
4 THE MODERN PERIODIC TABLE 52
5 LAYING THE FOUNDATIONS OF THE
   PERIODIC TABLE 68
6 THE FIRST NEW ELEMENT:
   THE DISCOVERY OF PHOSPHORUS 76
7 THE KEYBOARD OF THE ELEMENTS 86
8 GROUPS AND TRIADS 102
9 THINGS FALL APART 114
10 THE PERMANENCE AND IMPERMANENCE
   OF THE PERIODIC TABLE 130

APPENDIX 142
  BASIC GROUPS OF THE PERIODIC TABLE 144
  A SIMPLIFIED PERIODIC TABLE 154
  FURTHER READING 156
  INDEX 157
  ACKNOWLEDGEMENTS 160

1

DISCOVERING THE PERIODIC TABLE

SINCE EARLY CLASSICAL TIMES, NATURAL PHILOSOPHERS (as the early scientists were known) had believed that all matter was comprised of an admixture of four basic elements – earth, air, fire and water. This classification system persisted throughout the Middle Ages and into the Renaissance. However, over the course of the Enlightenment, thanks to the discoveries of the 18th-century French chemist Antoine Lavoisier, a new chemical classification emerged. Instead of four elements, there were now said to be 55 – each of which had the quality of being fundamental, in that it was irreducible to any simpler substances.

Subsequently, chemists began trying to classify these elements according to their groups of properties, but with little success. Meanwhile, other spheres of science had already made spectacular achievements. Copernicus and Kepler had begun to explain the solar system, placing the sun at its centre. Newton had described the workings of this system by means of gravity. And Darwin had proposed how life itself had evolved according to natural selection. The discovery of a pattern amongst the chemical elements might well uncover a missing link between physics and chemistry: a key to life on earth and the behaviour of the universe at large. Some believed this might even provide a blueprint to the cosmos. This discovery occurred in 1869 with the classification of the chemical elements in a periodic table.

The man responsible for this breakthrough was the Russian chemist and inventor Dmitri Ivanovich Mendeleyev (1834–

1907), born in a remote village near Tobolsk in Siberia, some 1,500 miles east of Moscow. He was the youngest of 14 or 17 children (no one seems quite sure, possibly owing to the number who did not survive childhood). His father had lost his teaching job when he went blind and had subsequently died when Dmitri was 13. This event had prompted Dmitri's forceful mother Mariya to re-open the family glass-making factory, but within a year this enterprise collapsed when the factory burnt down. Recognizing young Dmitri's outstanding academic talent, Mariya set off with him across Russia – travelling by foot, by cart, on horseback, by riverboat and by train – until finally they reached St Petersburg. Here she browbeat the authorities into letting the 16-year-old Mendeleyev study at the university, even though he had no relevant qualifications (Siberian educational certificates were not recognized in St Petersburg).

A year or so later, Mendeleyev's mother died and, overcome with grief and suffering from what would later be diagnosed as tuberculosis, her favourite son spent much of his remaining student life in bed. Illness and grief may have begotten physical idleness, but this did not extend to young Dmitri's mental capacities. He was still able to indulge his seemingly insatiable appetite for knowledge, and he graduated in 1854 at the top of his year; two years later, he was further awarded a master's degree in chemistry.

Mendeleyev spent a couple of years working for Robert Bunsen at Heidelberg University, and in 1860 attended an international

PRINCIPLES OF THE PERIODIC TABLE

The periodic table is an arrangement of all of the chemical elements, ordering them in vertical columns according to their group ('family') and across horizontal rows according to their periodic (regularly recurring) properties. Mendeleyev's fundamental insight states that, 'the elements, if arranged according to their atomic weights, exhibit an evident *periodicity* of properties'.

conference in Karlsruhe where significant time was devoted to the need to standardize chemistry, most notably by drawing up a classification of the elements. Mendeleyev returned to St Petersburg to teach in 1861 and was determined to improve the standard of chemistry teaching in Russia. In a typical frenzy of industry, he wrote a 500-page textbook in just 61 days. The result, *Organic Chemistry*, won him international recognition and the prestigious Demidov Prize (a sort of Russian precursor to the Nobel).

By the exceptionally early age of 32 Mendeleyev had become a professor of chemistry at the University of St Petersburg, his salary enabling him to acquire a small country estate at Tver, 300 miles south of the city. Here he conducted many personal agricultural and scientific research projects, as well as proposing progressive ideas in fields ranging through mineral exploration, spectroscopy, the elimination of smoke from the funnels of Russian naval ships, and petrochemical refining. (According to his early opinion, the use of petroleum as fuel was 'akin to firing a kitchen stove with banknotes.') Despite being a polyglot he retained a lifelong aversion to what he described as 'high culture', such as literature and the arts. This would make him a difficult father-in-law when his daughter Lyubov married the great lyrical poet Alexander Blok in 1903.

Mendeleyev's early academic fame enabled him to travel widely, delivering lectures throughout Europe and in America. After visiting the oil fields of Pennsylvania he changed his mind about petrol, and on his return home he set about transforming the fledgling Russian oil fields by introducing revolutionary new refining techniques and laying gas pipelines.

Despite such cosmopolitanism, Mendeleyev retained and even exaggerated certain characteristics to reflect his Siberian origins. Previous generations of his family were said to have contained a number of Kirgiz Tartars, and he liked to play this to the hilt, claiming to have been brought up by Tartars in eastern Siberia and to have spoken no Russian until he was 17. In support of such fibs, there was no denying Mendeleyev's eccentric appearance, which became more exotic the older he got. He had a long, unkempt beard that ended in distinct points, the result of his obsessively combing it with his fingers when deep in thought. His head supported a large, shaggy halo of unkempt hair, which he is rumoured only to have cut on an annual basis: with the coming of spring, he would summon a local shepherd from his estate who would set about his master's hair with sheep shears. Little wonder that when Mendeleyev visited London the Scottish chemist William Ramsay described him as 'a peculiar foreigner, every hair of whose head acted in independence of every other.'

A contemporary photograph shows Mendeleyev working at the desk of his study in St Petersburg. Behind him is a bookshelf with rows of bound volumes. Above his head, suspended from a hook on one of the shelves, hangs a key – looking much like a kind of scientific halo, or exclamation mark ('Eureka!')

NEWLANDS' LAW OF OCTAVES

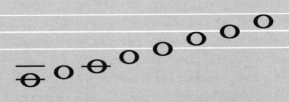

The English chemist John Alexander Reina Newlands (1837–1898) devised a periodic table of sorts some five years ahead of Mendeleyev, similar to the octaves in music. Newlands published his 'law of octaves' in 1865, arranging elements by their atomic number into seven groups where, 'any given element will exhibit analogous behaviour to the eighth element following it in the table.' Unlike the modern periodic table, periods were shown running from left to right and groups from top to bottom. However, the Society of Chemists declined to publish Newlands' finding, and many contemporaries dismissed his theory.

In Germany, too, a chemist called Julius Lothar von Meyer published an early periodic table in 1862, arranging 28 elements into six groups according to atomic weight and valency. (For an explanation of valency see page 46).

whilst, above the bookcase, high on the wall, hangs a row of framed portraits. Galileo, Descartes, Newton and Faraday – Mendeleyev's scientific Gods – gaze down at the wild-haired figure scribbling frantically amidst a disarray of papers scattered over his desk.

This apparent disorder is deceptive. What appear to be random scattered pieces of paper are in fact cards, each one headed with the name of an element, followed by its atomic weight and a list of its known properties, in which Mendeleyev was searching to discern a glimpse of some overall pattern.

ON MONDAY 17 FEBRUARY 1869, MENDELEYEV RECEIVED a visit from his colleague, Aleksandr Inostrantzvev. The household was paralyzed with dread: Mendeleyev – known for his frightful temper and outbursts – had ordered a troika to take him to the Moscow station (in Russia, stations tend to be named after their final destination) so that he could catch the train to his estate in Tver. However, he had reportedly been working for three whole night and days without sleep, and the servants were torn between fear of disturbing him and fear of him missing his train. With their master wrapped in thought, the maidservant dismissed the cabbie, and as the bells of his horse-drawn troika tinkled into the silence of the snowbound streets the manservant dragged Mendeleyev's heavy wooden trunk up the frozen steps into the front hall. And here they both stood, despondently awaiting their fate.

On the other side of the study door, Mendeleyev continued silently counting and comparing the properties of each of the known elements that he had written down on his cards. These were the fundamental chemical substances, which could not be further broken down into smaller parts that consisted of one or more other substances. Each had its own unique set of properties. When different elements combined, they formed molecules, and thus constituted the building blocks of the entire material world, both organic and inorganic. By the mid-19th century, dozens of these unique elements had been discovered, and more were being identified year by year. Their variety was both extreme and perplexing. Those already discovered included all manner of solids, gases and liquids. Many of them were capable of being transformed from one of these states to another, all the while remaining the same element. Iron, for instance, could be melted into a liquid. Yet contrary to the long-held belief of the alchemists, no element could be transformed into another element: base metals such as lead could not be transmuted into gold.

The variety of these elements – in both form and behaviour – was even more remarkable than alchemy. Some elements, such as gold, were malleable, but appeared able to withstand attack from the most corrosive of acids; others, such as carbon, could be found in forms ranging from crumbly black charcoal to brilliant, intensely durable diamond; where certain elements were extremely reactive, others were seemingly all but inert.

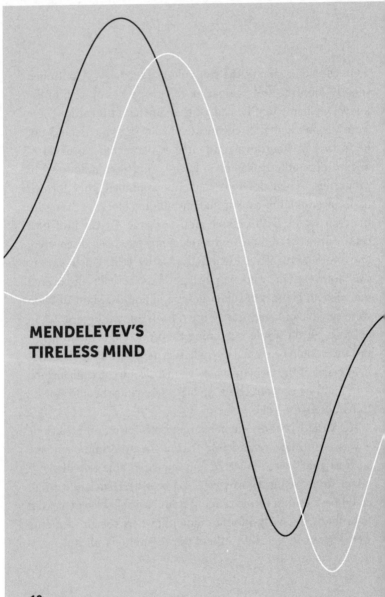

**MENDELEYEV'S
TIRELESS MIND**

Mendeleyev was known for prodigious bouts of concentration, during which he was capable of working for days on end. Equally, he could devote weeks, even months, to a particular project. Mendeleyev mentions in his diary how – after finding an early step towards making a discovery – he was liable to become manically elated; whilst later – on encountering difficulties – he would lapse into a deep depression, often bursting into tears of frustration. For Mendeleyev, chemistry was a passionate and dramatic activity.

The discovery of the periodic table was far from being Mendeleyev's only achievement: he proposed a number of practices to make farming more efficient, was a founder of the Russian Chemical Society, provided a definition for the absolute boiling point of a substance; and is credited with introducing the metric system to Russia.

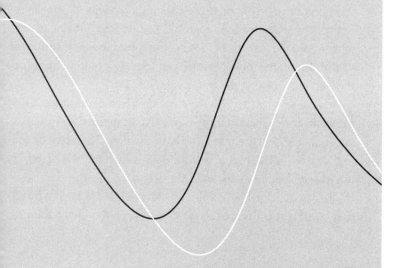

The discovery of these elements was transforming the way in which scientists understood the physical world: liquid water, long regarded as an archetypical element, was now known to be a molecule formed from hydrogen and oxygen, both of which occurred naturally as gases.

Mendeleyev was gripped by an obsessive belief that there must be a relationship between these elements, despite their disparate properties. It was now known that each element had its own unique weight. Surely there had to be some kind of pattern that ordered these weights in a coherent way. They could not, he believed, occur simply by accident. That said, ordering them according to their individual weights seemed to make no sense whatsoever, as neighbouring elements bore no resemblance to one another. Yet placing them according to their properties made no sense either. Collections of properties frequently overlapped, or matched in one way but not in any others. And the weights of those with similar properties differed wildly. No matter how Mendeleyev tried to group or number the elements, a coherent pattern eluded him.

Mendeleyev was in no doubt about the great significance of his research. If he could identify a pattern amongst the elements, one that classified them in an ordered fashion, this might be the first step to unlocking the pattern upon which life was based. It might even lead to discovering the origins of the universe.

This was no far-fetched dream. Mendeleyev was convinced that here lay the ultimate secret of God's creation. Even though he did not personally believe in God, he still thought in these terms, such was the ultimate significance of what he was doing. And he knew it. In scientific reality, he believed, there was no room for such thoughts as God, or divine powers, in the search for this ultimate secret. The elements required no such supernatural explanation. They were merely there, and they simply reacted with one another. What he was trying to discover was the pattern of how this took place.

Mendeleyev had studied the elements throughout his student years and, more recently, he had devised the idea of writing down the name of each element on a separate card – a notion that had come to him while playing his favourite game of patience on the long train journey to Tver. Now, sitting at his desk, the elements were as familiar to him as a class of pupils to an experienced schoolmaster. He knew all the telltale signs – the volatile ones, the easily influenced, the bullies, the gangs of miscreants, the dangerous loners. He had tried all manner of grouping them, but none of them worked. There just had to be a pattern somewhere: these were the building blocks of creation, and the foundation of the world simply could not be built upon an ad hoc scattering of elements.

Mendeleyev was certain that the sea of 63 labelled cards between them contained a similar pattern to a pack of playing cards. There were the different suits: hearts, clubs, diamonds and spades, within which numbers ascended in order; further variations were suggested by the royal cards and the aces.

KEKULÉ AND THE
BENZENE RING

Mendeleyev was not the first chemist to make an important discovery through a dream. A dozen or so years previously, the German chemist August Kekulé (1829–1896) had been pondering the complex structure of the benzene molecule, which was known to contain six hydrogen atoms and six carbon atoms. However, nobody could work out how these fitted together. Following a dream in which a snake swallowed its own tail, Kekulé realized that the carbon and hydrogen atoms could be joined in a circular structure (left) which became known as the benzene ring.

As he later wrote (with a mixture of optimism and hindsight) 'I knew it was all formed in my head, but somehow I just couldn't express it.' At last, overcome with fatigue, even Mendeleyev had to admit defeat. He leaned forward, resting his shaggy head on his forearms. Almost immediately, he fell into a deep sleep, during which, 'I saw in a dream a table where all the elements fell into place as required. Awakening, I immediately wrote it down on a piece of paper.'

Laying out his 63 cards as if they were a pack of playing cards, Mendeleyev created a table with each 'suit' forming a vertical column in ascending order through the numbers, the picture cards and the aces. Mendeleyev – still using the model of a pack of playing cards – then began to adapt the pattern, dividing his 63 cards into similar vertical columns, lined up beside each other. After sufficient juggling of the permutations, Mendeleyev arrived at an order in which many of the properties also seemed to match across their horizontal rows. Even so, they didn't fit completely, and there were even a number of gaps. Over the course of a morning, Mendeleyev sought to account for the discrepancies. The notes he made remain today, jotted down on the back of an envelope, which still bears the ring-mark from his mug of breakfast tea.

Mendeleyev brilliantly reasoned that the gaps must represent elements that had yet to be discovered. This was not unreasonable, given that new elements were being discovered on a regular basis. Furthermore, thanks to the table he could now even predict the weight and properties of these undiscovered elements. Finally, all the known (and even unknown) elements could be aligned in the columns and rows of a repeating periodic structure. For this reason, he decided to name his discovery The Periodic Table of the Elements.

At the time Mendeleyev formulated his periodic table, there were just 63 known elements. Now, the total number is 113, with the existence of several more (numbers 113, 115, 117 and 118) yet to be confirmed. Indeed, the American scientist Richard Feynman has suggested there may be as many as 137 elements. In honour of his achievement, Mendeleyev would eventually have an element named after him: Mendelevium (Md, element 101) is, appropriately, an unstable element, liable to spontaneous nuclear fission; it was discovered in 1955.

HOW THE PERIODIC TABLE WORKS

of the elements in a scientific paper in March 1869, and presented his findings to the Russian Chemical Society. As more elements, and even groups of elements, were discovered, his original table would undergo considerable modification. By examining the construction of his original table, however, the extent of Mendeleyev's achievement is revealed.

Mendeleyev's periodic table (*see* page 30) orders the elements sequentially to a clear set of criteria. Each element is made up of its own unique atoms. These can be distinguished by their properties, as well as by their differing atomic weights (the number of particles in the nucleus of one atom of a particular substance). The table starts with the top element of the first column on the left: this is hydrogen (H), which has an atomic weight of one. The elements are then listed in columns in numeric order by their ascending atomic weights (a feature that would greatly assist in suggesting the atomic weights of those elements that Mendeleyev predicted would fill the gaps in his table). The manner by which a column is determined – that is to say, the point at which one column ends and a new one begins – can at first glance appear almost arbitrary. However, the pattern becomes clear as soon as we start to read from left to right across the (horizontal) rows.

Take, for example, the seventh row from the bottom, which lists the four elements known as the halogens. All four elements – fluorine (F), chlorine (Cl), bromine (Br), iodine (I) – share certain resemblances or properties: they are all reactive, non-metallic elements with high electronegative qualities that make them the most reactive of all non-metals, especially so with alkali metals and alkaline earth elements where they then form stable crystalline substances. For example, chlorine (Cl) bonds with the alkali metal sodium (Na), to form sodium chloride, better known as common salt, (NaCl).

These shared chemical properties make them an unmistakable group, distinct from any other. Whilst their ascending atomic weights are very different – respectively 19, 35.6, 80 and 127 – they appear to be roughly double one another.

A similar pattern occurs in the row directly below the halogens, which contains the alkali metals (from left to right):

lithium (Li), sodium (Na), potassium (K), rubidium (Rb), caesium (Cs), and thallium (Tl)*

These, too, have particular shared characteristics: for metals, they are all surprisingly soft, and, when sliced, reveal a shiny inside that quickly tarnishes. This is because all these metals are highly reactive, even to air at room temperature. Indeed, they are so reactive that pure samples are stored in mineral oil. This high reactivity means that they are not discovered as pure elements in nature, but instead combine with other elements in the form of salts.

---

* Thallium is not listed in this group in the modern versions of the periodic table.

ОПЫТЪ СИСТЕМЫ ЭЛЕМЕНТОВЪ.

ОСНОВАННОЙ НА ИХЪ АТОМНОМЪ ВѢСѢ И ХИМИЧЕСКОМЪ СХОДСТВѢ.

			Ti = 50	Zr = 90	? = 180.
			V = 51	Nb = 94	Ta = 182.
			Cr = 52	Mo = 96	W = 186.
			Mn = 55	Rh = 104,4	Pt = 197,4.
			Fe = 56	Rn = 104,4	Ir = 198.
		Ni = Co = 59		Pl = 106,6	O- = 199.
H = 1			Cu = 63,4	Ag = 108	Hg = 200.
	Be = 9,4	Mg = 24	Zn = 65,2	Cd = 112	
	B = 11	Al = 27,4	? = 68	Ur = 116	Au = 197?
	C = 12	Si = 28	? = 70	Sn = 118	
	N = 14	P = 31	As = 75	Sb = 122	Bi = 210?
	O = 16	S = 32	Se = 79,4	Te = 128?	
	F = 19	Cl = 35,6	Br = 80	I = 127	
Li = 7	Na = 23	K = 39	Rb = 85,4	Cs = 133	Tl = 204.
		Ca = 40	Sr = 87,6	Ba = 137	Pb = 207.
		? = 45	Ce = 92		
		?Er = 56	La = 94		
		?Yt = 60	Di = 95		
		?In = 75,6	Th = 118?		

Д. Менделѣевъ

MENDELEYEV'S PERIODIC TABLE OF THE ELEMENTS

Mendeleyev published his table (left) in a paper in 1869. The title translates as *An attempted system of the elements based on their atomic weight and chemical analogies*, and is 'signed' D. Mendeleyev at the bottom.

As in sodium chloride, they are then liable to be stable. In their pure state, the alkali metals even react with pure water – with a tendency to increasing reactivity as their atomic weight increases (further right along the row).

Lithium (Li)** is the lightest of all the metals in the periodic table, and also the least dense: it readily reacts and is liable to catch fire when exposed to air. Progressing further along the row, pure caesium (Cs), for example, ignites spontaneously when exposed to air and explodes when plunged into water.

What leaps out from Mendeleyev's version of the periodic table are the 'gaps': there are no numbers (or elements) between hydrogen (H) whose weight is given as 1, and lithium (Li) with a weight of 7. Other numbers are not whole: the atomic weight of beryllium (Be) – at the top of the second column – is not 8 (as we might expect) or even 9, but the seemingly inexact value 9.4. Likewise, in the ascending atomic weights of the halogen group each of the elements has an exact number except for chlorine (Cl), which is assigned the number 35.6. These apparent discrepancies would not be resolved until the concept of atomic number, as distinct from the atomic weight, was introduced (*see* pages 56-7).

** The mood-stabilizing drug commonly known as lithium is in fact a lithium salt.

IN MENDELEYEV'S TIME, THE INTERNAL STRUCTURE OF the atom was unknown. Indeed, the atom was thought to live up to the original name it was given by the Greeks for the smallest possible particle: *atomos*, meaning 'uncuttable' or 'indivisible'. However, we now know that this is not true. Each stable atom consists of a positive nucleus, whose electric charge is neutralized by the negative electrons that orbit it. The nucleus consists of two types of sub-atomic particle: positive protons and electronically neutral neutrons.

The atomic *number* of an atom describes the number of protons in its nucleus. For instance, the chlorine atom has 17 positive protons, whose electric charge is neutralized by 17 circling electrons (of negative charge but negligible weight.) However, chlorine has an atomic *weight* (or mass) of approximately 35.5. This is accounted for by the fact that the most commonly occurring isotopes of chlorine are chlorine-35 (made up of 17 protons and 18 neutrons) and chlorine-37 (with 17 protons and 20 neutrons). These chlorine isotopes occur in nature approximately 75% and 25% of the time, respectively, and as such chlorine's atomic weight of 35.5 is calculated as:

$$[(75/100 \times 35)] + [(25/100) \times 37]$$

We can now see that the atomic weight of 35.6 ascribed to chlorine by Mendeleyev in his original table is not nearly as mistaken as might first appear.

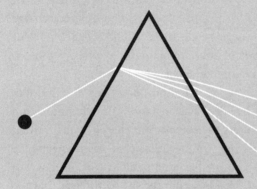

THE MARCH OF SCIENCE
IN THE 19TH CENTURY

In 1835 the French positivist philosopher Auguste Comte declared that science would eventually be able to discover everything that could be known, although he believed that there would be one or two exceptions, including the chemical composition of distant stars. The speed of scientific advances in the 19th century can be glimpsed by the fact that, thanks to advances in spectroscopy, just 14 years later scientists were able to analyse the light spectrum emitted by stars, from which they determined the chemicals of which they were composed.

The discrepancies in Mendeleyev's periodic table raised doubts within the scientific community about his findings, and to begin with his breakthrough was not widely recognized. Nonetheless, Mendeleyev's overarching idea would prove to be sound, especially with regard to the gaps in his rows and columns, and the properties of the elements that he predicted would fill them.

Take, for example, the ninth row down in Mendeleyev's table: this lists what became known as the boron group on account of its first element, boron (B). The next element in this row is aluminium (Al), following which this there is a gap, with a question mark and the atomic weight 68. Mendeleyev suggested that this gap would be filled by an element as yet undiscovered, with an atomic weight of 68 and characteristics in common with the other elements of the boron group. He even assigned a name to this element, calling it 'eka-aluminium', eka being the word for 'one' in the Sanskrit number system, and in this instance Mendeleyev meant that eka-aluminium was one place beyond aluminium in the group.

IN 1875, JUST SIX YEARS AFTER MENDELEYEV'S FIRST discovery of the periodic table, the French chemist Paul-Émile Lecoq de Boisbaudran (1838–1912) was using spectroscopy in an attempt to identify the impurities in some zinc ore that had been sent to him from the Pyrenees. Heating this ore under certain controlled conditions, Lecoq examined the lines emitted by the glowing ore within the light spectrum. Every single element was known to emit its own unique spectral lines (a process that has since been likened by modern scientists to its DNA). Lecoq noticed that his zinc ore sample was emitting spectral lines in a pattern that was hitherto unknown. After further experiments and analysis he finally isolated a new element, whose atomic weight he calculated as 69.9. He found that the new element was a metal with a colour like aluminium, with which it shared several other similar characteristics and properties. It is said that Lecoq was so proud of his discovery that he patriotically named it gallium (Ga), from *Gallia*, the Latin name for France. (The cockerel – *le coq* – is of course also the symbol for France, the Latin for cockerel being *gallus*. Some historians have claimed, therefore, that Lecoq in fact named gallium in his own honour.)

At the time of this discovery, Lecoq had not even heard of Mendeleyev's periodic table, let alone the prediction concerning eka-aluminium that Mendeleyev had suggested. When he was informed that his discovery had been predicted, Lecoq indignantly countered that gallium's atomic density (not to be

A GRAMMAR
OF THE
ELEMENTS

Mendeleyev correctly predicted that 'gaps' in his table would be filled by elements that had not yet been discovered. He gave these provisional names using the prefixes eka-, dvi- and tri- (the Sanskrit for one, two and three) depending on whether the undiscovered element was one, two or three places down from the known element with similar properties. Just as contemporary scholars – including Mendeleyev's friend, Otto von Böhtlingk – were uncovering the grammar of Sanskrit, so Mendeleyev was, in his own way, searching for a 'grammar' of the elements.

confused with its atomic weight ***) was very different from that predicted by Mendeleyev, thus proving that gallium could not possibly be eka-aluminium. However, as evidence gathered to support Mendeleyev's discovery and as scientists slowly grasped its revolutionary implications, Lecoq was proved wrong and Mendeleyev vindicated.

The discovery of gallium was the first, firm experimental evidence in support of Mendeleyev's table. Even so, many remained sceptical: there were simply too many gaps in his table. Take, for example, the elements in the row directly above the boron group: these begin with carbon (C) followed by silicon (Si), after which there is a gap (question mark) to indicate an element that Mendeleyev optimistically named 'eka-silicon'. For a further 11 years even Mendeleyev's supporters remained divided over the truth of some of his predictions and the overall accuracy of his table.

Finally, in 1886, a discovery by the German chemist Clemens Winkler at the University of Freiberg vindicated Mendeleyev and his periodic table. Winkler's research lead him to analyse a sample from a nearby mine of argyrodite, a compound known to contain sulphur and silver in the form of a silver sulphide. Winkler, however, discovered that this only accounted for

*** The atomic density of an atom refers to the number of atoms packed together in a unit of volume, whereas the atomic weight refers to the number of particles in the nucleus of each atom. Later calculations revealed the atomic density of gallium to be 5.9 g/cm3, almost exactly the number (6.0) predicted by Mendeleyev.

94% of its mass. The remaining 6% consisted of a compound containing sulphur and what appeared to be a previously undiscovered element, which he eventually managed to isolate. Echoing Lecoq's patriotism, Winkler decided to name this new element germanium (Ge).

On hearing about Winkler's discovery, Mendeleyev was convinced that this new element must belong to the eighth row of his table, the alkali earth metals – beryllium (Be), magnesium (Mg), zinc (Zn) and cadmium (Cd) – even though his periodic table did not anticipate any missing elements amongst this group. The properties of germanium pointed to its being eka-cadmium, Mendeleyev believed. This time, however, he was wrong. Germanium in fact belonged two rows below (but in the same column) next to selenium (Se) - rather than eka-cadmium it was eka-selenium. Ironically, Mendeleyev's initial misattribution of the element finally confirmed for many the veracity of his periodic table. The new element was only two places below eka-cadmium, and many of its characteristics were similar (as one would expect of elements in such proximity on the last column). Indeed, Mendeleyev had demonstrated that elements in the same column could also share certain properties. The discovery of germanium was too much to be a mere coincidence, and Mendeleyev's discovery was at last vindicated. His periodic table was recognized as a major advance in the scientific understanding of the nature of matter.

MENDELEYEV'S EIGHT PRINCIPLES

On 6 March 1869, Mendeleyev presented his paper *The Dependence between the Properties of the Atomic Weights of the Elements* to the Russian Chemical Society. In it, he described elements according to their atomic weight and the way they combine with other elements, stating eight main properties (some of which will be described in greater detail in the chapters that follow this one):

1. When arranged according to their atomic weight, elements exhibit an apparent periodicity of properties.

2. Elements with similar chemical properties have atomic weights that are either nearly the same – such as osmium (Os), iridium (Ir) and platinum (Pt) – or that increase regularly – such as potassium (K), rubidium (Rb) and caesium (Cs).

3. Arranging groups of elements in the order of their atomic weights corresponds to some extent to their distinctive chemical properties; as is apparent among groups such as lithium (Li),

beryllium (Be), boron (B), carbon (C), nitrogen (N), oxygen (O) and fluorine (F). There are also similarities in the way they combine with other elements.

4. The elements that are the most widely diffused have small atomic weights.

5. The magnitude of the atomic weight determines the character of the element, just as the magnitude of the molecule determines the character of a compound body.

6. We should expect the discovery of many yet-unknown elements – for example, two elements, analogous to aluminium and silicon, whose atomic weights would be between 65 and 75.

7. The atomic weight of an element may sometimes be amended by knowledge of those of its contiguous elements (the elements to either side).

8. Certain characteristic properties of elements can be foretold from their atomic weights.

3

HOW ATOMS REACT
WITH ONE ANOTHER

MENDELEYEV'S ORIGINAL PERIODIC TABLE, THOUGH revolutionary, would not survive unaltered: it would undergo considerable change to incorporate the discovery of additional, new elements.

First, however, it is necessary to understand one of the main ways in which atoms are able to work and react with one another. This is a concept called valency, which is the measure of an atom's ability to combine with other atoms and in what arrangement, and thus form a molecule (a compound of two or more atoms). The chemical formula for water, for instance, is H_2O: every molecule of water contains two hydrogen atoms linked to one oxygen atom. Imagine an atom to be a billiard ball with hooks attached to it, thus enabling it to link up with the hooks on other billiard balls. The number of hooks varies from atom to atom and is known as its valency. (Of course, atoms do not have actual hooks – the hooks are just a way to visualize the ability of atoms to react with one another.) A hydrogen atom has a valency of one (one hook), while an oxygen atom has a valency of two. This enables the oxygen atom to bond with the two hydrogen atoms to form water:

Similarly, the gas ammonia (NH_3) has one nitrogen atom of valency three, attached to three hydrogen atoms (each with a valency of one):

At the level of water and nitrogen, the concept of valency appears simple enough. Its complexity becomes apparent when we consider the challenge facing Kekulé as he tried to discover the atomic structure of benzene (*see* page 23). He knew that benzene consisted of six hydrogen atoms (each with a valency of one) and six carbon atoms (each with a valency of four). Each valency had to be linked to another (one hook attached to another), with no valencies left open (or hooks left unhooked). However, if each of the six hydrogen atoms linked to one of the six carbon atoms hydrogen, the molecule would be left with three unattached valencies:

H— C

The puzzle for Kekulé was how to fix six hydrogen atoms to six carbon atoms with their imbalance of valencies, without leaving a single valency unattached. As with so many insights of genius, the solution appears simple once we see it.

Kekulé's dream of a snake swallowing its tale inspired the idea of a ring of atoms – something hitherto inconceivable in molecular chemistry.

This gives a clear illustration of valency (from the Latin, *valentia*, meaning 'strength' or 'capacity' and hence, by extension the ability of an atom to combine with other atoms to form chemical compounds). When two atoms, or a group of atoms, combine to form a stable compound, all the valency links must be used up, with no link (or 'hook') left dangling. Atoms of the same element frequently combine, as in oxygen (O), and nitrogen (N), which occur in the air as gases in the form of O_2 and N_2. Elements within the main groups of the periodic table have a valency of between one and eight.

The benzene 'ring' discovered by Kekulé is a natural substance found in crude oil and is classified as organic chemistry. This is the branch of chemistry that deals with molecules, often highly complex in structure, containing carbon atoms. These are said to be organic – or in a loose sense, 'living' – and include substances from carbohydrates to DNA. Inorganic chemistry, on the other hand, is concerned with elements and compounds formed by combinations of any and all other elements. For centuries it was believed that many of the substances of organic chemistry contained some inherent 'life force', which distinguished them from 'dead' inorganic compounds. This was disproved in 1828 when the German chemist Friedrich Wöhler managed to create urea – an organic substance found in urine – from two inorganic substances, silver cyanate and ammonium chloride. Initially, it was claimed that Wöhler had succeeded in creating 'life'.

AN EARLY VERSION OF THE PERIODIC TABLE

I	II	III	IV	V	VI	VII	VIII		
H 1.01									
Li 6.94	Be 9.01	B 10.8	C 12.0	N 14.0	O 16.0	F 19.0			
Na 23.0	Mg 24.3	Al 27.0	Si 28.1	P 31.0	S 32.1	Cl 35.5			
K 39.1	Ca 40.1		Ti 47.9	V 50.9	Cr 52.0	Mn 54.9	Fe 55.9	Co 58.9	Ni 58.7
Cu 63.5	Zn 65.4			As 74.9	Se 79.0	Br 79.9			
Rb 85.5	Sr 87.6	Y 88.9	Zr 91.2	Nb 92.9	Mo 95.9		Ru 101	Rh 103	Pd 106
Ag 108	Cd 112	In 115	Sn 119	Sb 122	Te 128	I 127			
Ce 133	Ba 137	La 139		Ta 181	W 184		Os 194	Ir 192	Pt 195
Au 197	Hg 201	Ti 204	Pb 207	Bi 209					
			Th 232		U 238				

This early diagramatic form of Mendeleyev's table already shows refinements: compared to his original drawing (see page 30) we can see that it has been rotated clockwise through 90 degrees, with hydrogen (H) placed above lithium (Li). Also, to keep the atomic weights in the correct order (that is, ascending from left to right) the next six groups (II to VII) have simply been reversed. Group VIII resolves only partially some of the anomalies from Mendeleyev's original drawing. However, the principle of predicting elements that had not yet been discovered – as well as their probable weight and group properties – has been retained.

As further anomalies were resolved, this diagram, too, would evolve to become the modern periodic table.

THE MODERN
PERIODIC TABLE

IN 1894 THE SCOTTISH SCIENTIST WILLIAM RAMSAY AND the English physicist John William Strutt (Lord Rayleigh) made another discovery that would fundamentally change our understanding of chemistry. Rayleigh noticed that pure nitrogen (N) extracted from air (then thought to consist of approximately 78% nitrogen and 22% oxygen) had an atomic density of 1.257 grammes per litre. However, pure nitrogen obtained by chemical reaction from ammonia (trihydrogen nitride, or NH_3) had an atomic density of 1.251 grammes per litre. Following Rayleigh's observation, Ramsay devised an experiment to remove all the nitrogen from his sample obtained from air by passing the nitrogen over heated magnesium (Mg), which reacts with nitrogen to form magnesium nitride (Mg_3N_2). In doing so, Ramsay was left with a small amount of gas that had the curious property of not reacting with anything. It had a valency of zero! Spectral analysis revealed this to be a hitherto unknown element, which owing to its lack of reactivity Ramsay named argon (Ar) after the Greek αργον, meaning 'idle' or 'inactive'.

Given their extraordinary discovery, it is perhaps surprising that Ramsay and Rayleigh initially kept it a secret. However, the Smithsonian Institute in Washington had just announced a prize of $10,000 (the equivalent of at least $300,000 in 2015) for, 'some new...discovery about atmospheric air'. One condition of the prize was that the discovery should not have been announced before the end of 1894. In January 1895

Ramsay and Rayleigh duly announced their discovery and subsequently scooped the prize.

Ramsay and his assistants at University College, London, continued their research and, in 1898, by a process of fractional distillation they isolated three more elements and named them neon (Ne), krypton (Kr) and xenon (Xe), after the Greek νέος (*néos*, 'new'), κρυπτός (*kryptós*, 'hidden') and ξένος (*xénos*, 'stranger') respectively. Twelve years and a Nobel Prize later, in 1910, Ramsay obtained a pure sample of what proved to be another noble gas, helium (He), which had previously only been observed in the spectrum of the sun. (It was discovered in 1868 by astronomers Pierre Janssen and Joseph Lockyer whilst looking at layers of the sun's atmosphere, hence helium from the Greek ἥλιος (*helios*, 'sun')). That same year, Ramsay was the first person to isloate radon (Rn)*, a noble gas with a density of 9.7 grammes per litre, far greater than any other known gas.

Ramsay had now isolated no fewer than six noble gases: helium (He), with atomic number 2; neon (Ne), 10; argon (Ar), 18; krypton (Kr), 36; xenon (Xe), 54; and radon (Rn) 86. There was no doubting that this was a periodic group, but how could it be made to fit the table? The answer was to reorganize Mendeleyev's original (*see* page 30) to create a modified version (*see* pages 58-9). Indeed, as early as 1902 and the discovery of helium and argon, Mendeleyev had begun doing just this.

---

* Radon was actually discovered in 1900 by the physicist Friedrich Ernst Dorn.

ATOMIC NUMBERS
& ATOMIC MASS

The first three elements in the modern periodic table are hydrogen (H), helium (He) and lithium (Li), having atomic numbers 1, 2 and 3. These denote the number of protons in their nuclei. They were the first three elements to be created after the Big Bang, and all of the elements whose nuclei have higher numbers of protons would be created from these three original ingredients. This was done by the fusion of the smaller nuclei into larger ones in the extreme conditions of the early universe.

Every element has a unique atomic number, and in its uncharged state will have an equal number of electrons. The atomic *mass* of an element is determined by the number of protons and neutrons in its nucleus, and some elements have more than one atomic mass – this occurs when the number of neutrons varies, and is known as an isotope. Scientific convention denotes the atomic number as Z, from the German word *Zahl*, which means 'number', 'cipher' or 'figure'.

THE MODERN PERIODIC TABLE

GROUPS

	1	2		3	4	5	6	7	8
PERIODS									
1	HYDROGEN 1 **H** 1.0079								
2	LITHIUM 3 **Li** 6.941	BERYLLIUM 4 **Be** 9.0122							
3	SODIUM 11 **Na** 22.990	MAGNESIUM 12 **Mg** 24.305							
4	POTASSIUM 19 **K** 39.098	CALCIUM 20 **Ca** 40.078		SCANDIUM 21 **Sc** 44.956	TITATIUM 22 **Ti** 47.867	VANADIUM 23 **V** 50.942	CHROMIUM 24 **Cr** 51.996	MANGANESE 25 **Mn** 54.938	IRO 26 **F** 55.8
5	RUBIDIUM 37 **Rb** 85.468	STRONTIUM 38 **Sr** 87.62		YTTRIUM 39 **Y** 88.906	ZIRCONIUM 40 **Zr** 91.224	NIOBIUM 41 **Nb** 92.906	MOLYBDENUM 42 **Mo** 95.94	TECHNETIUM 43 **Tc** [98]	RUTHE 4 **R** 101
6	CAESIUM 55 **Cs** 132.91	BARIUM 56 **Ba** 137.33	57-70 *	LUTETIUM 71 **Lu** 174.97	HAFNIUM 72 **Hf** 178.49	TANTALUM 73 **Ta** 180.95	TUNGSTEN 74 **W** 183.84	RHENIUM 75 **Re** 186.21	OSM 7 **O** 190
7	FRANCIUM 87 **Fr** [223]	RADIUM 88 **Ra** [226]	89-102 **	LAWRENCIUM 103 **Lr** [262]	RUTHERFORDIUM 104 **Rf** [261]	DUBNIUM 105 **Db** [262]	SEABORGIUM 106 **Sg** [266]	BOHRIUM 107 **Bh** [264]	HASS 10 **H** [26

* LANTHANIDE SERIES	LANTHANUM 57 **La** 138.91	CERIUM 58 **Ce** 140.12	PRASEODYMIUM 59 **Pr** 140.91	NEODYMIUM 60 **Nd** 144.24	PROMETHIUM 61 **Pm** [145]	SAMA 6 **Sr** 150
** ACTINIDE SERIES	ACTINIUM 89 **Ac** [227]	THORIUM 90 **Th** 232.04	PROTACTINIUM 91 **Pa** 231.04	URANIUM 92 **U** 238.03	NEPTUNIUM 93 **Np** [237]	PLUTO 9 **P** [24

10	11	12	13	14	15	16	17	18
								HELIUM 2 **He** 4.0026
			BORON 5 **B** 10.811	CARBON 6 **C** 12.011	NITROGEN 7 **N** 14.007	OXYGEN 8 **O** 15.999	FLUORINE 9 **F** 18.998	NEON 10 **Ne** 20.180
			ALUMINIUM 13 **Al** 26.982	SILICON 14 **Si** 28.086	PHOSPHORUS 15 **P** 30.974	SULFUR 16 **S** 32.065	CHLORINE 17 **Cl** 35.453	ARGON 18 **Ar** 39.948
NICKEL 28 **Ni** 58.693	COPPER 29 **Cu** 63.546	ZINC 30 **Zn** 65.39	GALLIUM 31 **Ga** 69.723	GERMANIUM 32 **Ge** 72.61	ARSENIC 33 **As** 74.922	SELENIUM 34 **Se** 78.96	BROMINE 35 **Br** 79.904	KRYPTON 36 **Kr** 83.80
PALLADIUM 46 **Pd** 106.42	SILVER 47 **Ag** 107.87	CADMIUM 48 **Cd** 112.41	INDIUM 49 **In** 114.82	TIN 50 **Sn** 118.71	ANTIMONY 51 **Sb** 121.76	TELLURIUM 52 **Te** 127.60	IODINE 53 **I** 126.90	XENON 54 **Xe** 131.29
PLATINUM 78 **Pt** 196.08	GOLD 79 **Au** 196.97	MERCURY 80 **Hg** 200.59	THALLIUM 81 **Tl** 204.38	LEAD 82 **Pb** 207.2	BISMUTH 83 **Bi** 208.98	POLONIUM 84 **Po** [209]	ASTATINE 85 **At** [210]	RADON 86 **Rn** [222]
UNUNNILLIUM 110 **Uun** [271]	ROENTGENIUM 111 **Rg** [272]	COPERNICUM 112 **Cn** [285]	UNUNTRIUM 113 **Uut** [286]	UNUNQUADIUM 114 **Uuq** [289]				

GADOLINIUM 64 **Gd** 157.25	TERBIUM 65 **Tb** 158.93	DYSPROSIUM 66 **Dy** 162.50	HOLMIUM 67 **Ho** 164.93	ERBIUM 68 **Er** 167.26	THULIUM 69 **Tm** 168.93	YTTERBIUM 70 **Yb** 173.04
CURIUM 96 **Cm** [247]	BERKELIUM 97 **Bk** [247]	CALIFORNIUM 98 **Cf** [251]	EINSTEINIUM 99 **Es** [252]	FERMIUM 100 **Fm** [257]	MENDELEVIUM 101 **Md** [258]	NOBELIUM 102 **No** [259]

THE FIRST THING TO NOTICE ABOUT THIS REVISED TABLE is that the atomic numbers now read from left to right in rows, or periods, across the page. This is the case even when there is a significant gap, such as extends between hydrogen (H) with atomic number 1 (now raised to the top left where previously it was in one of the middle rows) and helium (He) with atomic number 2 (top right). Period 2 begins with lithium (Li), now found to have an atomic number of 3 **, followed by beryllium (Be), now found to have an atomic number of 4. These are followed by a large gap, before the series (period) continues with boron (B), atomic number 5, and ends with the noble gas neon (Ne), atomic number 10. Above and below neon (that is, running down the vertical column from helium (He) through neon (Ne) and all the way down to radon (Rn) in Period 6) is the group consisting of the other noble gases. In the column to their left (Group 17) we find the halogen group (*see* pages 28-29), including fluorine (F), chlorine (Cl), bromine (Br) and so forth – ending with astatine (At), although this would not be discovered until 1940.

Other groups in the right-hand half of the table include the copper group (sometimes also known as the coinage group), the zinc group, the boron group, the carbon group, the nitrogen group and the oxygen group (groups 11 to 16).

** Mendeleyev's figure had been 7, which is almost the exact atomic *weight* of lithium, whose nucleus consists of three protons and four neutrons.

Back at the beginning, Group 1 is headed by hydrogen (H) and followed by lithium (Li) and is usually known as the lithium group or as the alkali metals group. This brings us to another important property of the elements, which is their alkaline or acidic propensity. The word alkali derives from the Arab *al-qali*, meaning 'from ashes'. The elements in Group 1 are very reactive, and are thus usually found as compounds (that is, joined to other elements) often in the form of minerals. These minerals are usually extracted from the earth and smelted, after which the element arises 'from ashes'.

The elements of the 2nd group, headed by beryllium (Be), are often known as the alkali earth metals, because they can be found in pure form in nature, although owing to their comparatively high reactivity they, too, often occur as compounds. The liquid compounds of both groups are known as alkaline solutions. For instance, sodium (Na) and water (H_2O) can combine to form sodium hydroxide, a reaction that also releases the double-molecule of hydrogen gas (H_2):

$$2Na + 2H_2O \rightarrow 2NaOH + H_2$$

The prefix numbers (e.g., 2 in 2Na) represent the numbers of atoms required to achieve a particular reaction, enabling each of the different elements to combine according to its valency; it is also essential that the total numbers of each atomic element remain the same (otherwise matter would somehow have been

Owing to the predictive power of the periodic table, chemists had known for some years that there was a fifth member of the halogen group with an atomic number of 85. In 1931 the American physicist Fred Allison claimed to have discovered this, and called it alabamine (Ab) after his native state, Alabama. Six years later an Indian chemist made a similar claim. But neither substance fulfilled the predicted properties of element 85, and both were dismissed. The periodic table had proved an even more powerful tool than laboratory experiments. A convincing sample of element 85 – astatine (At) – was finally produced by a team of scientists at the University of California, Berkeley, in 1940.

THE PREDICTIVE POWER
OF THE PERIODIC TABLE

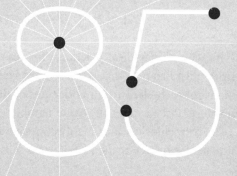

either destroyed or created). The final product in this instance is sodium hydroxide (NaOH):

This is a highly caustic compound, and liable to react with many substances with which it comes into contact – especially acids, which in chemical terms are the very opposite of alkalis (loosely comparable to the negative and positive charge in electricity).

The acids are formed from other groups in the periodic table. For example, the halogens (Group 17) can form acids (from the Latin *acidus*, meaning 'sour') when combined with water: chlorine gas (Cl_2) and water (H_2O) under certain conditions form hydrochloric acid (HCl), releasing a double molecule of oxygen gas:

$$2Cl_2 + 2H_2O \rightarrow 4HCl + O_2$$

As with all chemical reactions, for reasons of valency both sides must balance (that is, there must be the same amount of atoms of each element on both sides of the equation). In this instance, $2Cl_2$ denotes the same number of chlorine atoms as 4Cl. Hydrochloric acid is highly corrosive and, like most acids, it reacts readily with alkalis to form a salt. When hydrochloric acid and sodium hydroxide are mixed, they produce an extremely violent reaction, which forms common salt (NaCl) and water (H_2O):

$$HCl + NaOH \rightarrow NaCl + H_2O$$

The properties of such elements and their compounds, along with their reactions when mixed (such as acid and alkali to form a salt), have been known since earliest times. However, a full understanding of this would be reached only when the inner structure of the atom became known (*see* chapter 5).

Although the new periodic table contains a great number of changes and additions, it is clear that Mendeleyev's original idea remains at its heart: the elements are listed in order of their atomic number, in a table of periodically recurring groups of elements with similar, graduating properties. It is doubtful whether a definitive version of the periodic table will ever be found. There are several good, modern versions, each of which emphasizes different aspects of the recurring patterns and groups formed by the elements. The one shown on pages 58-9 is a version of the popular 'long form' table; a simplified version (with Groups 1-7 and 0) is commonly taught in schools (*see* pages 154-5). More ingenious versions include circular and helical versions; some are even three-dimensional. However, the fact that the elements form a pattern according to their periodicity remains a constant feature.

Nuclear fusion involves combining the nucleus of one atom with that of another, to form the nucleus of a different atom with a higher atomic number. It occurs only under extreme conditions. When atoms of a mass lower than iron combine, a huge quantity of energy is emitted. The sun generates energy by the nuclear fusion of hydrogen nuclei (one proton) into helium nuclei (two protons), which releases the vast quantities of light that enable it to illuminate the planets throughout the solar system. This process also generates sufficient heat to raise the temperature on its nearest planet, Mercury, to more than 400 degrees centigrade; to warm Earth, and even to raise the temperature of the furthest planets in the solar system.

The opposite of nuclear fusion is nuclear *fission*, which involves the division of atomic nuclei and causes a nuclear explosion, releasing a vast quantity of energy.

NUCLEAR ENERGY

5

LAYING THE FOUNDATIONS
OF THE PERIODIC TABLE

AS WE HAVE SEEN, EACH ELEMENT CONSISTS OF ITS OWN unique atoms, with their own distinct characteristics – atomic number, atomic weight, spectral 'signature', belonging to its own family group with quasi-similar properties, and so forth. Ironically, although some of these properties were already known, in 1869 when Mendeleyev proposed his periodic table most of the scientific community still did not even believe in the existence of atoms! The perfectly reasonable scientific argument was that no one had actually seen an atom, or provided any definitive evidence that such a thing existed. This state of affairs would continue for some years. Indeed, the matter was not finally settled until 1905, when Einstein published a paper demonstrating the existence of atoms, along with his papers on quantum theory and the special theory of relativity.

It is important to bear this scepticism in mind when considering the actual ingredients of the periodic table, the elements themselves. For despite any controversy over their nature – i.e., whether they consisted of atoms of not – the existence of elements had long been accepted.

The modern concept of an element, and its properties, had in fact been defined as early 1661 by one of the founders of modern chemistry, the Anglo-Irish scientist Robert Boyle (1627–91). In Boyle's historic words, elements were:

> *certain primitive and simple, or perfectly unmingled bodies; which not being made of any*

other bodies, or of one another, are the ingredients
of which all those called perfectly mixed bodies are
immediately compounded, and into which they
are ultimately resolved.

In plain terms, any substance that could not be broken down into simpler substances was an element. Boyle also understood that such elements could combine to form a compound ('clusters or groups'). Here was the first notion of what would develop into the modern idea of a molecule.

Unlike the Ancient Greeks – some of whom had proposed similar ideas – Boyle was not theorizing. Rather, he came to his conclusions as a result of persistent and sophisticated experimental work in his laboratory, and of the discoveries made by his contemporaries and predecessors under similar experimental conditions. This focus on *experiment* was the basis of modern science, and would prove hugely influential in the progress of modern chemistry. Boyle established the modern practice of performing particular actions with specific equipment in a listed order, using measured quantities. Most importantly, these instructions and their results were recorded and made available to fellow chemists, who could thus repeat the experiment. Experiments that did not produce the same results when repeated would then be discarded.

IN ADDITION TO POSITING THE IDEA OF THE ATOM, THE Ancient Greeks had also understood the notion of elements. However, for almost 2,000 years scientists had accepted Aristotle's (mistaken) belief that everything in the world was comprised of a combination of just four elements: earth, air, fire and water. Ironically, it was unorthodox learning – such as the 'hermetic' sciences and versions of pure 'magik' – that was not so constricted. However, such pursuits were frowned upon as hardly scientific, with the possible exception of alchemy, whose aim was the quasi-metaphysical task of turning base metals such as lead into gold.

To the modern mind, such alchemical projects may appear misguided, to say the least, but the *means* by which its practitioners attempted to achieve their aim were not. Experimental chemistry owes a profound debt to the alchemists, who in their futile pursuit of gold refined many of the experimental techniques – and much of the equipment – that would herald the birth of modern chemistry.

Having settled upon the definition of an element, Boyle and his fellow chemists quickly began taking note of resemblances between these elements and sought to classify them accordingly. This process was hampered in part by the fact that a number of compounds were mistakenly believed to be indivisible elements, simply because no method had yet been discovered of separating them. Even so, the main obstacle to a meaningful system of classification was that simply not enough elements (or even pseudo-elements) had yet been discovered for a clear pattern to emerge.

It is no accident that Robert Boyle himself, despite his modern scientific ideas, continued covertly to practise alchemy in his laboratory, as did Isaac Newton.

That they did so illustrates the limited understanding of matter in the 17th century. 'What was the 'secret' of matter?' The secret was that there was no secret. All matter was a combination of earth, air, fire and water – and, as such, it should have been possible to recombine the elements to form other substances. Indeed, there were known chemical processes that did just this, such as the formation of salts, distillation, fermentation, smelting ore and so forth. Alchemists – even Boyle and Newton – were more interested in transforming materials rather than investigating their ingredients. This was the categorical error that held back chemistry at exactly the time that physics, astronomy, mathematics and other pursuits were making huge progress and revolutionizing scientific thought.

THE ROYAL SOCIETY

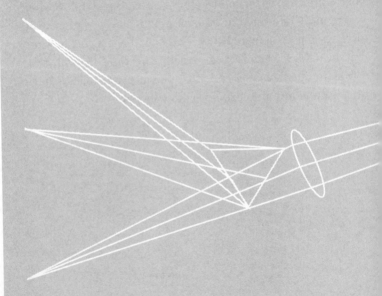

Robert Boyle (*see* page 70) was a member of the 1660 committee of 12, who are generally credited with founding the Royal Society of London, which would play a leading role in the progress of modern science. Its early members included such varied and illustrious figures as Isaac Newton, the physicist Robert Hooke, architect Christopher Wren and diarist Samuel Pepys. The Society's motto, '*Nullius in verba*', can be loosely translated as 'take nobody's word for it', a commitment by its Fellows to verifying scientific findings by experiment. Nonetheless, Boyle also indulged in some far-sighted scientific predictions, including the advent of manned flight, organ transplants and mind-altering drugs.

6

THE FIRST NEW ELEMENT: THE DISCOVERY OF PHOSPHORUS

WITHOUT QUITE KNOWING IT, THE ANCIENT GREEKS HAD in fact discovered ten elements. Not surprisingly, seven of these were metals: namely iron, copper, silver, tin, gold, lead and mercury, all of which (except for mercury) occur freely in nature. The other three elements were antimony, which has some metallic qualities; sulphur, whose fiery properties first brought it to notice; and arsenic, known for its efficacy as a poison. There followed, nonetheless, the long centuries when earth, air, fire and water were regarded as the only elements. That said, in about 1400 an unknown alchemist discovered bismuth, although he mistook it for lead. The name of this element comes from the German *Wismuth* – a corruption of *weiße Masse* meaning 'white matter' – although it was mistaken for lead until the middle of the 18th century.

No sooner had Robert Boyle's 1661 definition of an element gained acceptance within the new community of scientific chemists than the race was on to discover new elements. These were the first, defining steps towards the day when Mendeleyev would be able to lay out the 63 cards on his desk and articulate his theory of periodicity. The discovery of these new elements would assemble the components of what would become the periodic table, and they would also start to reveal the complex pattern of interwoven properties that would make the periodic table such a significant breakthrough. These early discoveries made possible the discovery of the periodic table, which would unlock the secrets of chemistry itself.

It took just six years from Boyle's definition for the first new element to be discovered. Ironically, it was made by an alchemist in pursuit of turning 'base matter' into gold (the fabled 'philosopher's stone'). The German alchemist, Hennig Brand (1630–1692), believed in the erroneous, ancient doctrine that nature revealed its secrets in symbolic form: for example, a natural substance that was gold in colour was liable to contain actual gold. With this in mind, Brand came up with the ingenious idea of experimenting on human urine, a practice so malodourous that it had previously deterred even the most intrepid alchemists from prolonged investigation.

Brand was nothing if not dogged in his investigations, collecting scores of barrels of his basic ingredients from miners in the nearby Harz mountains (who, in order to sustain them in their hot and thirsty work, drank prodigious quantities of beer). Brand then started on a series of experiments on his reservoir of accumulated urine – fermenting it, distilling it and otherwise subjecting it to trusted alchemical processes. He finally obtained a soft, whitish substance that glowed in the dark and was liable to spontaneous combustion unless kept under water. On account of these exceptional properties he named this substance phosphorus (*phos*, Greek for 'light'; and *phoros*, 'bringing'). However, in true alchemists' style, Brand kept the secret of how to manufacture phosphorus to himself – earning a small fortune from demonstrations of its wondrous qualities at royal courts throughout Germany.

The shared characteristics between phosphorus and the other elements in the nitrogen group result from their internal atomic structure (*see* page 82). Suffice to say that the atoms of all of the elements in this group have five electrons in their outer shell (an outer orbit of electrons around the central nucleus) and each element has one more shell than the preceding element (nitrogen has two, phosphorus has three, arsenic has four and so on). These electrons determine the chemical behaviour – the properties – of the atom itself. The outer orbit is known as the valence shell, which in turn determines the valency of the atom (the 'hooks on the billiard ball', *see* page 46). Under normal circumstances, all the elements in Group 15 have a valency of five.

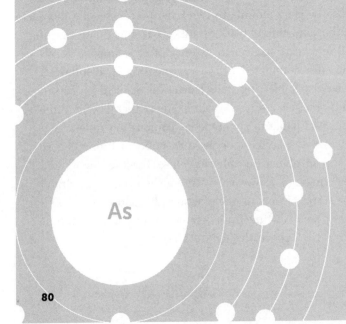

As

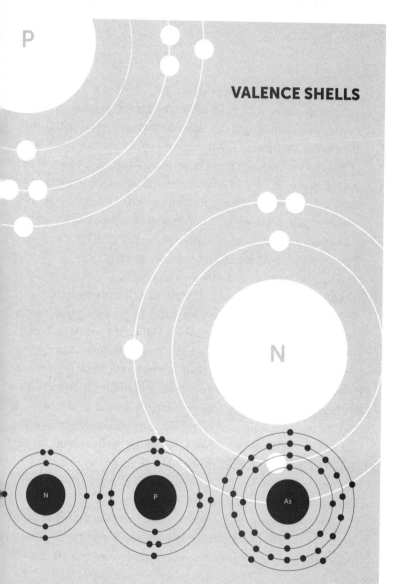

P

N

WITH HINDSIGHT, PHOSPHORUS CAN BE SEEN TO SHARE certain qualities with bismuth as well as with arsenic, which occurs as a whitish-grey metalloid (that is, a semi-metal with properties somewhere between those of metals and non-metals). Two hundred years later, phosphorus (P), arsenic (As) and bismuth (Bi) would be grouped as three of the five elements forming Group 15 (the nitrogen group), whose properties include an increasingly metallic character as the atomic number increases, as well as a similarly increasing melting point and boiling point, and decreasing electronegativity.

With the exception of nitrogen (N) itself, which is a gas and sits at the top of the group, they all occur as increasingly heavy, light-coloured solids. Nitrogen itself becomes a solid under extreme conditions: first, at minus 196°C, it becomes a liquid; then, at minus 210°C, the liquid nitrogen freezes to become a crystalline solid. (The reverse of this process can be seen in ice, which melts at 0°C to become water, and in turn boils at 100°C to become steam). Crystalline nitrogen would prove to be a highly reactive solid and to exhibit unmistakable similarities to those elements in Group 15 whose solid form occurs comparatively naturally. Even early on after the discovery of phosphorus, however, it was possible to glimpse how it might fall into place with arsenic and bismuth.

As we have seen, phosphorus is so reactive that it cannot occur in nature in its pure state; rather, it occurs in various minerals, such as phosphates, whose molecules consist in part of phosphorus atoms. Phosphorus has some very important commercial applications, especially in the form of fertilizers. This is because phosphorus is such an important, indeed vital, element in animal and plant life. Farm animals. which are not able to roam freely and find sources of phosphorus for their diet, must be given it as a supplement in their feed. Similarly, when plants are harvested, phosphorus is removed from the soil and must be replaced by fertlizer, otherwise the soil becomes barren. Thanks to agri-business, the production of organophosphorus has become a global industry.

The use of phosphorus in matches has largely declined, but it occurs in many other commerical products, from baking powder to toothpaste, soft drinks to preserved meats and cheese. Phosphorus is also a component in the special glass used for sodium lamps, and is highly valued for its miliary applications, which include incendiary bombs and tracer bullets.

In spite of Hennig Brand's best efforts, the secret of how to prepare phosphorus soon leaked out. When Robert Boyle obtained a sample, he investigated its properties in his laboratory, and he quickly succeeded in extracting phosphorus from human urine. Soon, phosphorus was being widely used in the preparation of a number of more or less efficacious (and more or less dangerous) medical preparations. Such was the demand for the new element that one of Boyle's laboratory assistants, Ambrose Godfrey, set up in business to manufacture phosphorus and made a fortune selling it all over Europe.

COMMERCIAL PRODUCTION OF PHOSPHORUS

In the 18th century, the Swedish pharmacist Carl Scheele's discovery of how to extract phosphorus from horse bones initiated the mass-production of this element, which led directly to the invention of the phosphorus match. It is no accident that Sweden became the world's leading producer of matches, culminating a century later in the worldwide business empire of Ivar Kreuger, 'the Swedish match king'. At the height of his success, Kreuger produced a third of the world's matches and his personal fortune was so great that he occasionally even subsidized national governments. (Until, that is, the Wall Street Crash of 1929 heralded the Great Depression and with it the collapse of Kreuger's business interests, personal bankruptcy and eventual suicide.)

7

THE KEYBOARD OF
THE ELEMENTS

OVER THE COURSE OF THE 17TH AND 18TH CENTURIES, the richness and variety of different elements were gradually revealed, as well as some of the patterns in which they combined. The new modern chemistry was soon developing a variety of innovative techniques, casting off the cloak of obfuscation bequeathed by alchemy. In feeling their way towards the emergent structure that would one day form the periodic table, chemists were also unwittingly indicating how its theoretical structure would in time come to suggest practical experiments.

About a century after Hennig Brand's discovery of phosphorus, Carl Scheele (1742–1786) contrived a method for preparing phosphorus from bones. Although elected to the Royal Swedish Academy of Sciences, Scheele spent most of his adult life as a pharmacist in the small town of Köping, carrying out experiments at the back of his shop. In the 15 years before his death at the age of 43, he published almost 50 scientific papers and discovered more elements than anyone before or since. The techniques he developed revealed a vast range of chemical compounds, and gave the first indication of how a scientific table of elements could be exploited. Scheele may not have discovered the periodic table but his experiments heralded a golden age for chemists. As I once wrote:

> *The undiscovered elements lay before them like the notes of a piano keyboard. They could play a few of these elementary notes, as well as a number of*

compound chords. But as their hands explored the keyboard, they gradually became aware for the first time of the vast range of tonal possibilities which lay before them. Scheele extended the possibilities of this keyboard more than any in his time.

In addition to his discovery of a means to mass-produce phosphorus, Scheele was the first person to isolate such elements as chlorine (Cl), molybdenum (Mo), manganese (Mn) and barium (Ba). All of these would play vital roles in new processes that heralded the Industrial Revolution: chlorine was used for bleaching textiles, for example, and manganese was alloyed to steel to make railway tracks that were both stronger and also less susceptible to rust and corrosion. But Scheele's greatest claim to fame was one of his earliest discoveries – of the gas that would become known as oxygen (O). As a result of this discovery, chemists would overturn their entire conception of the very air we breathe and how we breathe it (by taking in oxygen). It would also transform their understanding of the way many major chemical reactions work, disproving the long-held theory of phlogiston, which had been holding back modern chemistry for more than a century.*

---

* The term phlogiston first appeared in 1667 and was thought to be an element that was released into the air when a substance burned. In fact, the opposite occurs: burning takes oxygen *out* of the air. This 'oxidation' process would soon be recognized as one of the most widespread and natural chemical reactions, integral to the rusting of metals, the decay of living substances and the breaking down of cell structures.

CARL SCHEELE'S UNACKNOWLEDGED GENIUS

The biochemist and science-fiction writer Isaac Asimov has called Scheele the 'hard-luck' chemist, because so many of the Swede's discoveries have been credited to other scientists: for example, Scheele discovered oxygen, but the English chemist Joseph Priestley published a paper about it first; and although Scheele discovered chlorine, he mistook it for a compound, and it was another 35 years before Humphry Davy, another English chemist, identified and named it as an element. Scheele's ultimate hard luck was his untimely death at the age of 43, poisoned by his scrupulous habit of personally smelling and tasting the properties of his experiments.

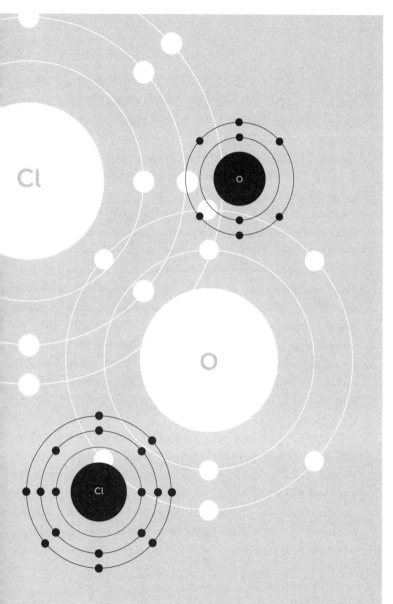

MANY MORE EXCITING DISCOVERIES WERE MADE OVER the course of the 18th century, as chemical experiments became more sophisticated and Western explorers ventured into territories previously unknown to Europeans. Just a few of these will suffice to indicate their range and the role they played in expanding scientists' knowledge both of the elements and of the patterns of their behaviour.

In 1735, a French sailor was walking along the shore of the Rio Pinto in western South America when he came across some spongy lumps of greyish clay that looked like cannon balls. Inside was found a new greyish metal, which was called *platina del Pinto* ('little silver of the River Pinto'), eventually becoming known as platinum. Although spongy in its natural state, when hammered and compressed this new metal proved as malleable as gold, and its resistance to chemical corrosion was found to be even greater than that of the world's most valuable metal. Platinum was soon being made into jewellery, but more importantly it was used for experimental equipment involving the strongest acids.

Another important discovery was made closer to home. The philosopher and scientist Henry Cavendish was reportedly so reclusive that he ordered his servants to hide whenever he emerged from the laboratory where he worked in his large London home. Despite this pathological shyness, he felt obliged to attend meetings of the Royal Society to pass on his findings, which were eccentrically conveyed in words interspersed with squeaks and little cries. However, his experimental expertise was without parallel, and in 1766 he achieved his greatest triumph with the discovery of the gas that would be called hydrogen**. In subsequent experiments he would prove that hydrogen and oxygen were the constituents of water – which, since the earliest civilizations, had been considered as an indivisible element. Decisively, hydrogen was found to have an atomic weight of one – later understood to indicate that its nucleus consisted of one proton circled by just one electron. No atom could be simpler.

Hydrogen would prove to be 'the first element of Creation' and the most numerous in the universe: the element out of which all others had been created after the Big Bang. Hydrogen fuels the sun, the stars and supernova by means of a process of nuclear fusion (*see* page 66), where, under conditions of immense gravitational pressure, vast amounts of energy are released (in the form of starlight, sunlight, heat) and the one-proton nuclei of hydrogen atoms fuse to form larger nuclei, which themselves fuse to become even heavier elements.

IN 1789, MARTIN KLAPROTH AT THE UNIVERSITY OF BERLIN discovered an element at the opposite end of the atomic scale from hydrogen. This new element would later be found to have an atomic number of 92 and an atomic weight (protons

---

** The name hydrogen comes from the Greek ὑδρο (*hydro*, meaning 'water') and γενής (*genes*, meaning 'forming') and was coined by Antoine Lavoisier in 1783.

plus neutrons) of 238, making it the heaviest element yet discovered. (We now know that to balance the positive charge of the protons requires no fewer than seven shells – or orbits – to accommodate its 92 negatively charged electrons.) Klaproth named his newly discovered element uranium, after the newly discovered planet Uranus.

Atomic nuclei can appear in variant forms known as isotopes, as typically exhibited in uranium. Different isotopes of the same element share the same number of orbiting electrons, and the same number of protons in their nucleus, but will include different numbers of neutrons. This means that they retain the same basic elemental properties, but their nuclei will have differing sizes and, often, differing stability. The simplest isotopes are those of hydrogen. Under normal circumstances, hydrogen has a nucleus consisting of one positive proton, which is orbited by one negative electron. However, hydrogen also occurs as two rarer isotopes: deuterium (with a nucleus containing one extra neutron) and tritium (which has two extra neutrons):

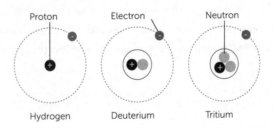

These isotopes are usually named after the size of their nucleus: so, in this instance, as hydrogen-1, hydrogen-2 and hydrogen-3 (H1, H2 and H3 for short).

Isotopes occurring in uranium are more complicated. The element that Klaproth discovered was in fact uranium-238. However, it was also found to have an isotope with three less protons, thus known as uranium-235. The lighter U235 is less stable than U238, and is subject to radioactive decay. As it decays, U235 emits energy in the form of subatomic particles and rays, before splitting into smaller atomic nuclei, usually krypton (Kr) and barium (Ba), the most common results of atomic fission of U235. Harnessing this 'atom-splitting' process on a massive scale would lead to the atomic bomb. However, it would be another 150 years before this lethal process was finally realized. Back in the 18th century, the sciences remained largely separate, although the chemistry of Boyle was at last beginning to catch up with the physics of Newton. Indeed, scientific disciplines were advancing at such a rate that the need for classification was increasingly pressing to keep pace.

SUPERNOVAE

A supernova is a stellar explosion of such immense force that it can outshine even a galaxy. These explosions can last for weeks on end and – by a process of nuclear fusion on a grand scale – emit more energy than the sun will during its entire lifetime. Stellar explosions are so powerful that they can fuse together larger nuclei to form heavier elements, which they scatter through the universe. The first recorded supernova in the Milky Way was observed by Chinese astronomers in 185 AD; and the most recent (in the Milky Way) was Kepler's Star in 1604.

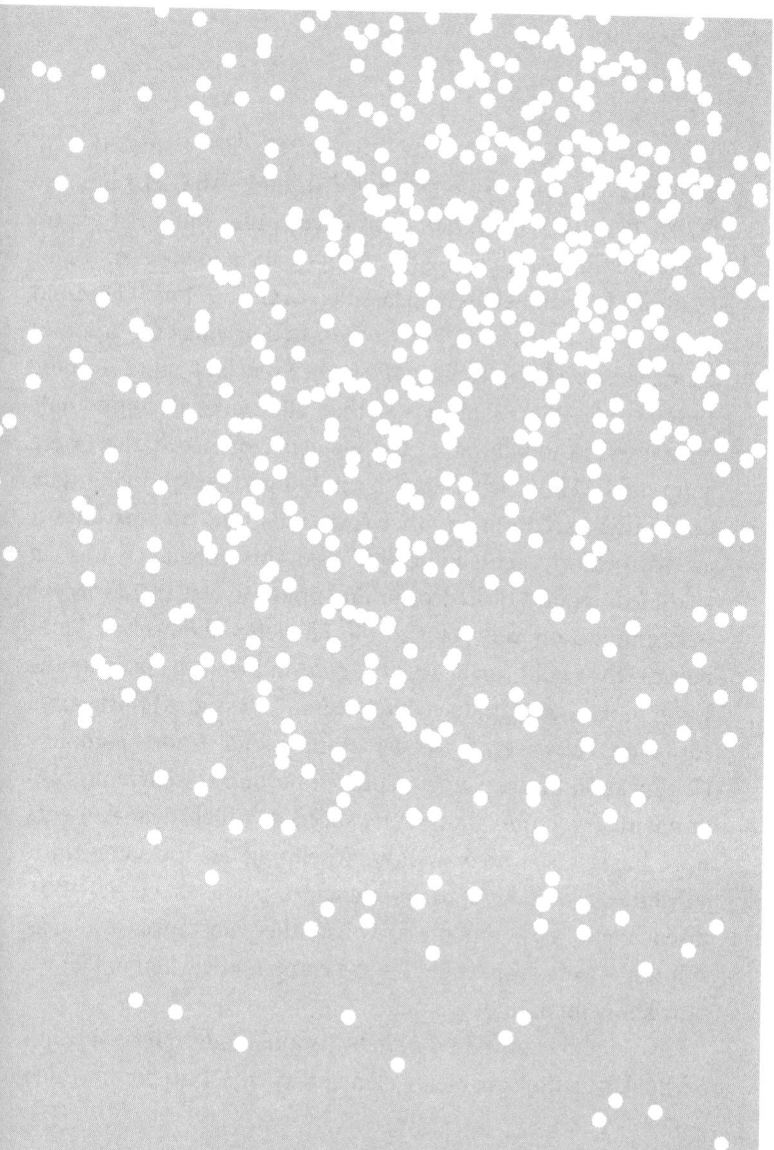

THE FIRST GREAT ADVANCE IN THE FIELD OF classification came from the Swedish botanist Carl Linnaeus (1707–78) who in 1753 introduced his binomial Latin nomenclature for all living things by genus and species: so, for example, human beings are classified as genus '*homo*' (hominid or primate), species '*sapiens*' (wise). In conscious imitation of Linnaeus, two decades later the great French chemist Antoine Lavoisier (1707–78) proposed that all chemical compounds be named rationally after the elements of which they were composed. Thus, what had been known as 'spirits of salt' (and other variations in other countries) now became known as hydrochloric acid, indicating that this substance was an acidic compound of hydrogen and chlorine (HCl). Lavoisier's next innovation was to give the elements themselves Latin names: thus gold (*or* in French, *guld* in Swedish) became the Latin *aurum*, and in due course would be assigned the symbol Au. Here, however, Lavoisier stopped: any more complete classification of the elements proved beyond the current grasp of chemistry. Quite simply, the speed at which new elements were being discovered was overwhelming. At the same time, substances long held to be elements (such as water) were discovered to be compounds, whilst other substances – such as sulphur – were suspected of being compounds when in fact we now know them to be elements.

Lavoisier also proposed a slightly more modern modification of Boyle's definition of an element as 'the last point which analysis is capable of reaching'. Yet the difficulty in actually pinpointing an element remained. Without a comprehensive system of classification, identifying types of elements and predicting ways they would react with one another remained very much on a trial and error basis. A system of *ad hoc* rules had accumulated over centuries of experiments: acids reacted with alkalines to become salts (although a few, for some reason, did not). Most solids melted to become liquids, which often vapourized to become gases (but again, some appeared not to do so). Some substances burned, others did not. And so forth. But in spite of this hard-earned knowledge, any discernable underlying structure remained elusive. In this respect at least, chemistry still remained a matter of educated guesswork.

What could possibly be the link between such disparate elements as phosphorus, hydrogen and uranium, let alone the other known elements? Was it possible that they just existed – as had previously been thought about earth, air, fire and water – only with more complex properties and wider-ranging differences? Surely this would be the equivalent of mathematics consisting of numbers that had no underlying relationship with one another: you could add them, even subtract them, but there would be no prime numbers, no factors, no strictly delineated building blocks upon which to construct a rational understanding that went beyond mere rule of thumb.

HYDROGEN

Hydrogen (H) is the lightest element in the periodic table, and has an atomic number of one. This is because it contains just one proton and one electron. It is also the most commonly occurring element in the universe, making up 75% of known matter and representing 90% of all atoms. In the aftermath of the Big Bang, as the universe expanded and cooled, electrons and protons were amongst the first particles formed. These positive and negative particles came together to form neutral hydrogen.

Hydrogen is a colourless, odourless, tasteless and non-toxic gas that was named by the French chemist, Antoine Lavoisier, in 1783 from the Greek *hydro* (ὑδρο) meaning 'water' and *genes* (γενής) meaning 'creator'. However, Robert Boyle discovered that the gas could be produced through the reaction of iron filings and acids in 1671, whilst Henry Cavendish is generally credited with its discovery as an element 100 years later. Hydrogen-filled balloons offered mankind the first safe and reliable experience of travelling by air, although such use was put paid by the *Hindenburg* disaster in 1937, which was erroneously ascribed to a hydrogen leak.

8

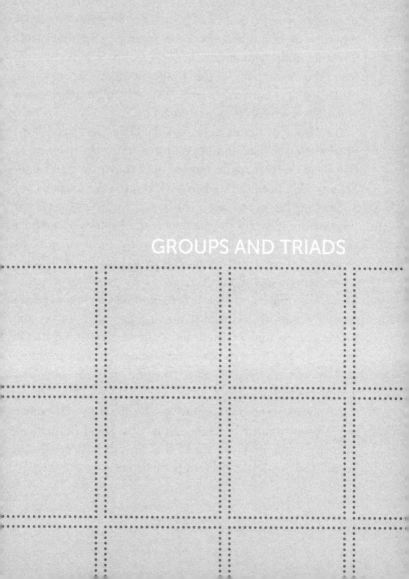

GROUPS AND TRIADS

WHEN MENDELEYEV LAID OUT HIS 63 CARDS IN ORDER of atomic weight, and recognized that certain properties repeated at identifiable periodic intervals, the first truly scientific understanding of chemical groups was born. Here were the equivalent of the prime numbers amongst the ascending list of atomic weights. This was the structure upon which the chemical elements were built.

The layout of the modern periodic table (*see* pages 58-9) divides the elements into 18 groups, of which the noble gases (Group 18) are perhaps the most obvious and easily identifiable: helium (He), neon (Ne), argon (Ar), krypton (Kr), xenon (Xe) and radon (Rn), whose apparent lack of properties (inertness) accounts for why they did not appear in Mendeleyev's original table. Initially Mendeleyev did not accept their existence as elements. Indeed, even 16 years after the publication of his table, when Ramsay was arguing the case for the existence of argon, Mendeleyev published an article suggesting that argon, whose atomic weight was claimed to be 40, was in fact a triatomic molecule consisting of three atoms of nitrogen, whose atomic weight was 14. Given contemporary inexactitudes, as exhibited in Mendeleyev's original table (*see* page 30), the opposing arguments both had merit.

What won the day for Ramsay was the discovery of the other noble gases, and their similar properties. Later, the pattern of their atomic numbers would confirm the existence of the group. Looking at the modern periodic table we can see that their atomic numbers are 2, 10, 18, 36, 54 and 86. The first three elements are separated by eight, the number of elements in both the second and third periods. The fourth period has 18 elements, so the next element down in the noble group has an atomic number 18 higher than the previous element. This pattern is repeated in the fifth period, which also has 18 elements. (After this, it becomes more complicated.) However, once again it was the periodic table itself that confirmed the existence of the group, whose characteristics conformed to their predicted properties. Under standard conditions the noble gases are all odourless, colourless, monatomic gases – that is, they appear naturally as single atoms, unlike, say, oxygen or hydrogen, which appear naturally as diatomic molecules (O_2, H_2). The atoms that form the elements of the noble group each have a completed outer shell of electrons, leaving no electron free to react (or hook up) with other atoms or molecules – thus they have a valency of zero. And as their atomic mass increases, so does their boiling point.

Precisely because of their inactivity, the noble gases have several practical and commercial uses. Argon is used in light bulbs with tungsten filaments, which can thus achieve great heat without risk of a reaction; it is also used to package food such as crisps, pizzas and salads to increase their shelf life by replacing the oxygen that would otherwise cause decay. Inert gases can also be pumped into the empty tanks of oil and gastankers to prevent residual flammable gas from exploding.

Research has shown that helium also has a number of unexpected properties. For instance, in 1937 it was discovered that liquid helium (cooled to minus 271°C, close to absolute zero) becomes a 'superfluid', which has zero resistance to flow and zero viscosity (basically 'thickness', in the sense that treacle is thicker than water). This means that at such a temperature superfluid helium even defies gravity: it will flow up the side of a beaker and even up over a partition. The full potential of superfluidity has yet to be developed, but already liquid helium has found applications in quantum research and in its ability to reduce the speed of light.

AT THE OTHER SIDE OF THE TABLE WE HAVE THE ALKALI metals of Group 1 (*see* pages 29, 32 for a summary of their characteristics). Even before Mendeleyev spotted a comprehensive, periodic pattern, a relationship between these elements had been formulated that would provide insights and prove crucial to the development of the periodic table. This was the 'triadic theory' of the elements, proposed in 1829 by the German chemist Johann Döbereiner (*see* pages 112–113).

Döbereiner's theory linked trios of neighbouring elements into 'triads', where the intermediate (middle) element has chemical properties and an atomic weight that are the approximate average of the two on either side of it. In Group 1 of the modern periodic table, lithium (Li), sodium (Na) and potassium (K) are all soft, reactive metals; they form a triad, where the atomic weight of sodium (11) is nearly the average of the atomic weight of lithium and potassium (3 and 19). Lithium reacts with water in a mild manner that releases comparatively little energy, whereas potassium is violently reactive; and sodium – the intermediate element – exhibits a reaction somewhere between the two. All three react to create solutions that are alkaline. Similar patterns occur in other properties, such as reacting with chlorine to form the diatomic molecules lithium chloride (LiCl), common salt (NaCl) and potassium chloride (KCl); or combining in a similar fashion with hydrogen (lithium hydride, LiH etc.), and with hydrogen and oxygen to form hydroxides: lithium hydroxide (LiOH), sodium hydroxide (NaOH, or caustic soda) and potassium hydroxide (KOH). Caustic soda can inflict serious burns and is usually handled in solution. It has widespread industrial uses, from detergents and drain-cleaner to aluminium production and upgrading crude oil. It is also used in the production of harder soaps, whilst a softer soap can be obtained with potassium hydroxide; and completing the triadic relationship, there is also a non-corrosive (intermediate) lithium soap.

Among the halogens in Group 17, we find that chlorine (Cl), bromine (Br) and iodine (I) also form a triad. Their atomic weights are 17, 35 and 53 respectively, and they are all powerful-smelling, poisonous gases with their own distinct colour. In solution they become acidic (e.g., hydrochloric acid, HCl). These acidic solutions readily react with the

ABSOLUTE ZERO

Absolute zero is minus 273.15°C. This is the lowest temperature possible, at which a substance has no heat whatsoever. Since heat is generated by the vibration of molecules or atoms, to all intents and purposes all movement ceases at this temperature. As yet, absolute zero has never been attained under laboratory conditions, although in 1999 an experiment at Helsinki Technology University reached a temperature of minus 273.14999999999°C (so, within 0.000000001°C of absolute zero). The lowest *natural* temperature ever encountered is minus 272.15°C, which was recorded in 2003 by astronomers observing the Boomerang Nebula, part of the Centaurus constellation which contains Alpha Centauri, the nearest star to our solar system (some 5,000 light years away).

alkali solutions of the lithium group (above) to form salts. For example:

$$HCl + NaOH \rightarrow NaCl + H_2O$$

(that is, hydrochloric acid mixed with sodium hydroxide → sodium chloride and water, commonly known as saltwater).

Triads can also straddle three groups in the same period (i.e., along one of the horizontal rows of the periodic table). This is the case in the second period, with oxygen (atomic weight 16), nitrogen (14) and carbon (12). However, apart from their weights, Döbereiner himself failed to find any other convincing links between this triad, assuming that, having identified the triad, others would soon discover chemical resemblances.

Some suggested groups turned out not to be triads after all: silver (Ag) in Group 11, Period 5, was thought to form a triad with mercury (Hg) in Group 12, Period 6, whose intermediate element was lead (Pb) in Group 14, Period 6. The three elements shared unmistakable triadic resemblances: all were metals of a silvery colour, whilst between the solidity of silver and the fluidity of mercury lay malleable lead. Yet it soon became clear that despite such resemblances, this was in fact no triad.

It is easy for us to 'read off' triads from a modern periodic table, but of course Döbereiner and his followers had no such guide. All they had was an incomplete list of known elements, their atomic weights (many of which would prove unreliable) and their limited knowledge of properties associated with the 54 elements that had been identified at the time.

In all, Döbereiner and his followers would identify nine triads, but they were limited in their research by the fact that simply not enough elements were known. Even so, Döbereiner's 'law of triads' constituted a considerable achievement, and would prove the first decisive step towards Mendeleyev's breakthrough. To this day, many scientists remain convinced that there is a triadic relationship that lurks within the several versions of the modern periodic table and its groups – and several hundred triads have been suggested, although their triadic resemblances are often based on highly differing evidence.

Indeed, no less an authority than Eric Scerri of UCLA, a leading philosopher of chemistry, insists:

> *I don't think we can yet settle the issue of the definitive periodic table or whether the left-step table (a variant on the 'standard' modern table) is superior to a triadic motivated table.*

That said, *most* chemists now concur that Mendeleyev's idea of periodicity is incontrovertibly the superior key to the elements and their properties.

DÖBEREINER
THE AUTODIDACT

H							He
Li	Be	B	C	N	O	F	Ne
Na	Mg	Al	Si	P	S	Cl	Ar
K	Ca	Ga	Ge	As	Se	Br	Kr
Rb	Sr	In	Sn	Sb	Te	I	Xe
Cs	Ba	Tl	Pb	Bi	Po	At	Rn

112

Johann Döbereiner (1780–1849) was the son of a coachman and largely self-educated. Yet such was his enthusiasm for learning that he was apprenticed to an apothecary, and in due course began publishing articles based on his chemical experiments. Eventually, Döbereiner worked his way up to becoming a professor of chemistry at the University of Jena, Germany where he formulated his theory of triads, an early precursor to the periodic table. Döbereiner's lectures proved so inspiring that they were attended by his friend the poet Goethe, who for years after this devoted much of his imaginative energies into scientific speculation – even coming up with the theory that all plants developed from an ur-species, a precursor to evolution.

9

THINGS FALL APART

IN THE YEARS FOLLOWING ITS PUBLICATION IN 1869, Mendeleyev's periodic table gained increasing recognition. Modifications to incorporate newly discovered elements (many of them predicted by the gaps in Mendeleyev's layout) and an increasing acceptance of the idea of the atom served to consolidate his reputation and the acceptance of the table. Yet even as this was happening, discoveries were being made that threatened the very notion of the atom as the fundamental particle of matter. Ironically, these very developments would eventually deepen our comprehension of the periodic table and our present-day recognition of all that it stands for.

The first doubts seemed innocuous enough. In 1895 the German scientist Wilhelm Roentgen (1845–1923), working at the University of Würzburg, noticed something that seemed capable of escaping from a lightproof box and causing a shimmering effect on a screen that happened to be at the other end of his laboratory. He called this an x-ray, after the common algebraic notion 'x' for an unknown. Experimenting with these x-rays, he took a picture of his wife, which revealed her skeleton, causing her to exclaim, 'I have seen my death.' This discovery of previously unknown rays received confirmation when the French scientist Henri Becquerel (1852–1908) discovered the phenomenon, later known as radioactivity, where a number of different rays were emitted by uranium. Consequent to this discovery, the Polish-born scientist Marie Curie (1867–1934) and her French husband, Pierre (1859–1906) undertook further investigations of this phenomenon in their small laboratory in Paris, and discovered that radioactivity was emitted by uranium and similar elements such as thorium. In the course of these experiments, the Curies would discover two new heavy elements both of similar atomic weight – polonium, named after her native Poland, and radium. These, too, produced a seemingly constant emission of radiation. How could this be? What was going on? Madame Curie wrote that, 'every atom of a radioactive body functions as a constant source of energy'. Not surprisingly, when the press got wind of this some years later, newspapers declared: 'Curies Discover Perpetual Motion!' At the very least, it looked as if a new atomic property had been discovered in elements that had a high atomic weight.

In an entirely separate development, the British physicist Joseph John (J. J.) Thomson (1865-1940) conducted a series of experiments at the Cavendish Laboratory, Cambridge that would lead to the discovery of the electron in 1897. The electron was indubitably a sub-atomic particle – in other words, a distinct part of an atom – and was also found to have a negative charge. The consequences of this discovery were truly groundbreaking: the atom was not 'uncuttable', as the ancient Greeks had believed, but had an internal structure of component parts.

Thompson explained that it looked as if each different elemental atom was like a spherical 'cake' with a positive electrical charge, which was embedded with enough electrons (which Thompson likened to raisins) to neutralize this charge:

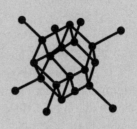

THE SWEDISH ACADEMY

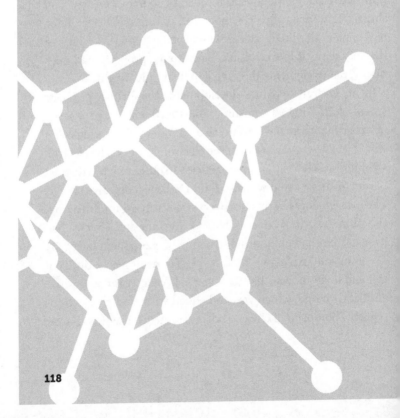

In 1903 Marie Curie became the first woman to receive the Nobel Prize for her work on radioactivity (although originally the Academy planned only to make the award to her husband, Pierre, and the physicist Henri Becquerel). Eight years later, she would be the first person to receive a second Nobel prize, this time in chemistry for her discovery of radium and polonium (named after her native Poland). Marie Curie died in 1934, aged 66, from aplastic anaemia, which it is thought she contracted from her long-term exposure to radiation. She remains the only woman to have won the Nobel Prize twice. Her daughter, Irène Joliot-Curie, won the Nobel Prize in chemistry along with her husband for the discovery of artificial radioactivity in 1935.

Whilst almost all of the great modern scientists have won Nobel Prizes, Dmitri Mendeleyev, ironically, did not. He was nominated for the Nobel Prize in chemistry in 1906, and it was thought that he was a shoo-in: however, two members of the influential Nobel Committee – Peter Klason and a rival of Mendeleyev's, the chemist Svante Arrhenius – argued that the periodic table was 'too old' for recognition and proposed instead the Frenchman Henri Moissan, who won for his work in isolating fluorine from its compounds.

thus the discovery of electrons provided another clue to explain how the periodic table worked.

One of Thomson's research students at the Cavendish was the bluff New Zealander Ernest Rutherford (1871–1937), whom he described as 'a charming blend of boy, man and genius.' Rutherford concurred with Thomson's picture of the atom, but remained puzzled by the phenomenon of radioactivity. After nine years spent at McGill University in Montreal, Rutherford returned to the UK in 1907 and, at the University of Manchester, began a new set of studies with his German assistant, Hans Geiger (who later invented the Geiger counter, for measuring radioactivity). These experimental studies led Rutherford to two important conclusions. First, it looked as if radioactivity – which included the emission of alpha particles, beta particles and gamma rays – somehow came from *within* the atom. Secondly, instead of Thomson's somewhat unscientific 'cake' analogy, he began to conceive of the atom as a sphere of positive gas studded with negative electrons. In order to investigate this, he devised an ingenious experiment. In it, he made a sheet of atoms (gold foil hammered to such thinness it was just a few atoms thick) that he bombarded with a stream of alpha particles (that is, two protons and two neutrons that are bound together and that are identical to a helium nucleus). These alpha particles passing through the gold sheet would register on the other side as tiny points of light on a fluorescent screen. Rutherford was not surprised that most of the alpha particles simply passed through the gold foil without deviation as, he assumed, the atoms consisted almost entirely of gas. Occasionally an alpha particle would experience a slight deflection, of around 1°: this, Rutherford reasoned, evidently occurred when it struck an electron. However, he was astonished when he saw that a few of the alpha particles were deflected by as much as 10°. Then, to his utter amazement, one alpha particle actually bounced *back* from the gold sheet. The only way Rutherford could account for this phenomenon was if the atom had a solid nucleus. In that moment, the science of nuclear physics was born.

Later, 20th-century nuclear physicists would discover that electrons circled the nucleus in a number of different orbits. Put simply, the inner orbit was capable of containing two electrons, the next two orbits could contain up to eight electrons before they were filled, the next could contain up to 18, the next 32 and so on, with the number of electrons increasing with the number of shells.* When an outer orbit of electrons is full, the atom does not have any 'free' electrons to react with other atoms, giving it a valency of zero. Now we can see how the atomic numbers of the noble gases indicate their inertness. (Their atomic number derives from the number of protons in their nucleus, and thus

---

* Interpretations differ: I have retained the older orbital configuration, which makes it easier to explain valency. An idea of the complexities of quantum mechanics can be seen in the fact that the electron may be viewed as a continuous wave, which encircles the nucleus in a wide 'smear', at any point in which the electron has only a probability of existing.

THE CAVENDISH LABORATORY

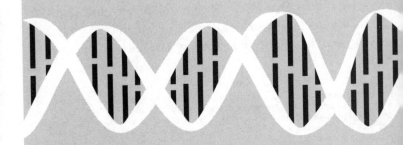

Contrary to popular belief, the Cavendish Laboratory at Cambridge University is not named after the famous 18th-century chemist Henry Cavendish, but after William Cavendish, seventh Duke of Devonshire and Chancellor of the University from 1861–91, who provided its funding. Over the years since it opened in 1874, the Cavendish has witnessed some of the greatest scientific developments of the modern era. In addition to J. J. Thomson's discovery of the electron (which led to the discovery of quantum mechanics in the 1920s), it was here that James Clerk Maxwell, the Laboratory's first director, formulated the theory of electromagnetic radiation (1862–4), which inspired Einstein's theory of relativity. Later on, it would see the invention of the cloud chamber by Charles Wilson, used to detect ionizing radiation; Rutherford's discovery of the atomic nucleus (*see* pages 120-1), and the structure of the double helix molecule of DNA by Francis Crick and James Watson.

the same number of negative electrons circling the orbits around this nucleus.) So, the noble gases have two protons and two electrons (He), then 10 (Ne), 18 (Ar), 36 (Kr), 54 (Xe) and 86 (Rn). In helium the first orbit is full; in neon the second orbit; in argon the third and so on (*see* pages 128–9). This pattern recurs throughout the periodic table. Now we can see why. In Group 17 – the Halogens – we have fluorine (F) with an atomic number of 9 (that is, nine protons, thus nine electrons), which means its inner orbit is filled with two electrons, while its outer orbit is incomplete, having only seven. Next, chlorine (Cl) has 17 electrons: its two inner orbits are filled (with two and eight electrons), again leaving seven electrons in its outer orbit. This pattern is repeated through the group: it is the *structural* reason why the halogens each have a valency of seven.

THE PUZZLE OF RADIOACTIVITY AND ITS LITERALLY earth-shattering consequences would take longer to solve. In 1938, the German chemist Otto Hahn (1879–1968) conducted an experiment that bombarded uranium with nuclear particles in the hope of creating a larger nucleus, and of discovering a new element heavier than uranium, then the heaviest known, naturally occurring element with atomic number 92.

Hahn and his co-worker, Lise Meitner (1878–1968), were regarded as the world's finest experimental chemists at the time, but their partnership was broken up when Meitner was forced to flee to Sweden owing to Nazi persecution of the Jews. So

Hahn continued the experiments on his own, which seemed to produce a transuranic element that bore a curious resemblance to the rare earth element lanthanum, as well as a residue of barium. Although he appeared to have created a new element, he also appeared to have performed the impossible alchemical transmutation of uranium into the toxic element barium in Group 2 (the alkali earth metals, headed by beryllium).

Hahn secretly communicated this result to Meitner. It was Meitner who understood what had happened: the uranium atoms had split into atoms of smaller nuclear numbers. Hahn had discovered nuclear fission, which involves the splitting of atomic nuclei and which in part involves the destruction of mass and subsequent release of energy according to Einstein's formula $E = mc^2$. Here E is energy, m is mass, and c is the speed of light (a mind-boggling 186,000 miles per second) – indicating that a tiny amount of mass is capable of becoming an *immense* amount of energy. The discovery that nuclear fission was actually possible (and not just purely theoretical) was kept a secret and smuggled to America where the Italian-American physicist Enrico Fermi used it in experiments that led to the first atomic bomb. This was a direct consequence of the periodic table. In attempting to create the larger nuclei of new transuranic elements, scientists had achieved 'the impossible' of splitting an atom. By destroying elements they had created vast amounts of energy – a process that could be used for bombs, or controlled and harnessed for the production of nuclear energy.

TRANSURANIC
ELEMENTS

In the mid-20th century, scientists believed that elements with an atomic number greater than 92 (known as transuranic elements) occurred when uranium atoms were bombarded with protons. A number of scientists claimed to have created elements 93 and 94, including Irène Joliot-Curie and Enrico Fermi, who went on to build the world's first nuclear reactor at Chicago University in 1941. The puzzle of radioactivity and the prospect of extending the periodic table to include new, artificially created elements had gripped the imagination of the top scientists of the time. In due course, these new elements would be created and, in recognition of their early efforts, element 100 was named fermium (Fm) and element 105 joliotium (Jl) although the latter is now known as dubnium (Db), after Dubna, the town in Russia where it was first produced in 1968.

ATOMIC STRUCTURE

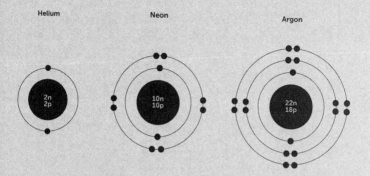

Helium   Neon   Argon

Every atom consists of a positively charged nucleus (consisting of positive protons and uncharged neutrons) around which revolve negatively charged electrons. Protons, neutrons and electrons are collectively known as subatomic particles, and it is the unique combination of these particles that gives an element its distinctive characteristics.

The atomic number of an element derives from the number of protons in its nucleus, with the same number of electrons circling it. When the first orbit fills, so a second one is formed, as can be seen with the noble gases.

10

THE PERMANENCE
AND IMPERMANENCE
OF THE PERIODIC TABLE

NUCLEAR FISSION NOTWITHSTANDING, SCIENTISTS HAVE, in fact, been more intent upon creating new elements than in breaking down known ones. As it happened, the first artificially produced element was not a transuranic, though it was created in the very years that Fermi, Joliot-Curie and Hahn were conducting their competing experiments aimed at the heavier elements predicted by the periodic table. The creation of the new element in question is a textbook example of Mendeleyev's original predictions.

For years, there had been a tantalizing gap in the periodic table where element 43 should have appeared. Its properties could even be predicted: it was in Group 7, the transition metals, of which all the known members had a valency of two (with two electrons in their outer shell). The transitions metals of Group 7 are headed by manganese (Mn), whose alloy is used in steel for strengthening prison bars and railway tracks. (Manganese is also one of the earliest elements to have been put to creative human use: Stone Age people used its oxide for cave paintings.) Directly below the 'gap' left by the missing element 43 was the metal rhenium (Re; in Group 7, Period 6), which is alloyed with nickel to strengthen airplane engine parts and turbine blades. Where rhenium has an atomic weight of 186 (75 protons and 111 neutrons), one of its isotopes has an atomic weight of 188, with a half-life of just 17 hours. This means that it takes 17 hours on average for half a quantity of rhenium-188 to decay, during which its unstable nuclei emit radiation until it transforms into a more stable state – a nucleus with a more stable arrangement of protons and neutrons – or splits into different nuclei altogether, which contain lower numbers of protons and neutrons by means of fission.

The missing element 43 – known as technetium (Tc), after the Greek *teckhnetos* for 'artificial' – was created by Fermi's colleague and friend Emilio Segrè as a result of cyclotron experiments in which molybdenum (atomic number 42) was subjected to a high-speed barrage of deuterons (protons and neutrons), resulting in nuclear fusion forming a nucleus containing 43 protons. As predicted by its place in the periodic table, technetium turned out to be a heavy metal, with a number of different isotopes of short half-life. The reason it had not previously been discovered was that the technetium present at the creation of the earth had – like so many other isotopes – decayed over millions of years, this nuclear fission resulting in other elements of different atomic weight. As if in confirmation of its place amongst the elements, technetium also appears to be the third member of a triad in the fifth period formed by niobium (Nb; atomic weight 92.9); molybdenum (Mo; weight 95.9) and technetium (Tc; weight 98.9). The weight of molybdenum is midway between the other two in the triad, whose members share several similar properties. For instance, all have numerous radioactive isotopes, with half-lifes [sic] of varying length, and some isotopes of niobium and molybdenum decaying into isotopes of technetium.

Not all half-lifes are as brief as those of technetium's isotopes. For instance, the isotope carbon-14 (which has two more protons than the commonly occurring, stable isotope carbon-12) has a half-life of 5,370 ± 40 years. Exploitation of this astonishing fact has enabled the science of carbon dating, which is used in archaeological fields ranging from the dating of dinosaur bones and ancient stone tombs to paint pigments in the identification of art forgeries. Carbon dating was famously used on the Turin Shroud, which was said to bear an authentic image of Jesus Christ until carbon dating proved that the cloth dated from the 1300s. It was also used to date Ötzi Man, whose mummified body was found, perfectly preserved, amongst melting glacier ice high in the Alps in 1991. Carbon dating was used to establish that he had lived, and died, around 3,300 BC. Examination of his carbon tattoos showed that they masked spots on his body that suggest acupuncture had been practised (at least 2,000 years earlier than its first known use in China).

THE USE OF OTHER RADIOACTIVE ELEMENTS AND THEIR calculated half-life (known as radiometric dating) has enabled us to reach even further into the past. Potassium-40, for instance, has a stupendous half-life of 1.3 billion years, during which it decays into argon-40, enabling archaeologists to measure the relative proportions of these isotopes and thus date rocks over one billion years old. Radiometric carbon dating has enabled scientists to establish that the oldest rocks in the solar system are some 4.6 billion years old. Such techniques are being used at this very moment by landing craft on Mars and the moons of Saturn and Jupiter, where they will reveal further secrets of the universe.

Not surprisingly, the half-life of the transuranic elements, with their large and unstable nuclei, is often very short. Meitnerium (Mt), named after Lise Meitner, has an atomic number of 109 and an atomic weight of 278. It also has seven known isotopes, ranging from meitnerium-266 to -279. Meitnerium-279 has the longest half-life, which is just six minutes. Experiments have detected traces of what appear to be other isotopes of meitnerium that exist for mere microseconds, and these are thought to be part of the decay sequence of even heavier elements. All this makes it virtually impossible to determine all but a few of the chemical and physical properties of such fleeting elements. (We can, of course, predict their likely properties from their location in the periodic table.)

The incredible short half-lifes of some of these isotopes raise the intriguing question, is there a limit to the periodic table? What is the highest atomic number possible? The highest named element 112 is copernicium (Cn), named after the great Polish astronomer Nicolaus Copernicus who suggested a model of the universe with the sun, rather than the earth, as its centre. However, unconfirmed (and unnamed) elements range as high as element 127, which is known as unbiseptium (Ubs; makeshift Latin for 'one-two-seven-ium').

RADIOACTIVE DECAY AND HALF-LIFES

Radioactive atoms have unstable nuclei, and the process by which they break down and change into a different atom altogether is known as radioactive decay. Although it is not possible to predict *when* an atom might decay, it *is* possible to measure the time is takes for half of the nucleus of a radioactive atom to decay – that is, its half-life. Different radioactive isotopes have very different half-lifes, ranging from thousands of years in the case of carbon-14 to a matter of minutes for the seven isotopes of meitnerium. Radioactivity decreases with time: suppose a substance has a half-life of 24 hours: in the first 24 hours it will decay by half (from 100 to 50); in the second 24 hours by another half (from 50-25) and so on for subsequent 24-hour periods.

The great 20th-century quantum physicist Richard Feynman, speculated that there could even be elements as high as 137 but no higher. This number is related to the fundamental fine-structure constant (α) – which defines the strength of electromagnetic interaction – and is linked to the charge of the electron, whose value is widely accepted as 1/137.

Many of these recently discovered elements have been named after famous scientists. These include einsteinium (Es), rutherfordium (Rf), roentgenium (Rg) and mendelevium (Md). The element platinum (Pt) was originally called platina, after *platina del Pinto* ('little silver of the River Pinto'), but the idea of a feminine name for a metal would prove anathema to the Victorian scientific establishment, who held sway in the naming of elements and changed it to its present name. All elements discovered since shortly after Queen Victoria's accession to the throne have been given the Latin neuter suffix *–ium*, or the Greek neuter *–on*, as in the case of the noble gases. This sexless nomenclature even extends to curium (Cm), named after Madame Curie. The sole exception to this rule is astatine (At), which has a female ending derived from the Greek *astatos*, meaning 'unstable'. (One can only speculate as to how this reflects on the predominantly male society of chemists.)

Regardless of the antiquated naming practice, the periodic table is assured of universal immortality. In the words of the celebrated science writer, John Emsley:

As long as chemistry is studied there will be a periodic table. And even if someday we communicate with another part of the universe, we can be sure that one thing that both cultures will have in common is an ordered system of the elements that will be instantly recognizable by both intelligent life forms.

CARBON

Carbon is the chemical basis for all known life (organic chemistry). By weight, it is the second most abundant element in the human body (after oxygen), the fourth most abundant in the universe, and the fifteenth most abundant in the earth's crust. Carbon commonly exists in more than one form (known as allotropes) – the best known are graphite and diamonds. It is also a major component of carbonate rocks, including limestone, marble and coal.

All living plants and animals contain carbon. When they die, the amount of the radioactive isotope carbon-14 starts to decay, with a half-life of $5,370 \pm 40$ years. Measuring the amount of carbon-14 in a dead plant or animal can reliably date objects from the last 50,000 years or so.

APPENDIX

BASIC GROUPS OF THE PERIODIC TABLE

Hydrogen (H): the periodic table includes this element theoretically as part of Group 1. However, it has such different properties that it is usually considered on its own. Hydrogen is the lightest and most abundant of all elements in the universe, and accounts for almost 90% of all atoms.

Group 1 – The Alkali Metals: lithium (Li), sodium (Na), potassium (K), rubidium (Rb), caesium (Cs) and francium (Fr)

Like all periodic groups, these have ascending atomic numbers (and weights). The alkali metals have strong similarities: though solid, they are soft at room temperature, and shiny when kept in a neutral substance such as oil. When exposed to air they become increasingly reactive. All occur in nature: sodium and potassium are the most abundant; fancium the least.

Group 2 – The Alkali Earth Metals: beryllium (Be), magnesium (Mg), calcium (Ca), strontium (Sr), barium (Ba) and radium (Ra)

Less reactive than the alkali metals – because of two electrons in their outer shell, which are more difficult to dislodge – these gained the name 'earth' metals during the Middle Ages because they do not decompose when heated. Being denser than Group 1 metals, they are better conductors of electricity.

THE TRANSITION METALS – GROUPS 3-12

There are 38 transition metals across groups 3 to 12 (sometimes known as the d-block of the period table).

Group 3 – The Transition Metals: scandium (Sc), yttrium (Y), lutetium (Lu) and lawrencium (Lr)

These metals, whose members have resemblances to elements in other groups, make it a controversial group. For instance, yttrium is sometimes regarded as a rare earth element. Like all metals they are ductile (i.e., they can be drawn out into wire), malleable and can conduct heat and electricity. Scandium was one of the elements to fill a 'gap' left by Mendeleyev in his original table. Lawrencium is a rare and unstable transuranic with a half-life of one hour 35 minutes, and is named after American physicist Ernest Lawrence, inventor of the cyclotron. The transition metals occur rarely in the earth's crust: like scandium, they are usually found in trace amounts in minerals.

Group 4 – The Transition Metals : titanium (Ti), zirconium (Zr), hafnium (H) and rutherfordium (Rf)

Both titanium and zirconium were discovered in the late 18th century by the German chemist Martin Klaproth. These transition metals are usually found in mineral form. Hafnium is used in nuclear reactors (to slow down the reaction) and in microprocessor chips.

Group 5 – The Transition Metals: vanadium (V), niobium (Nb), tantalum (Ta) and dubnium (Db)

The first three metals in this group, in ascending weight, tend to be hard, not easily worn down, and resistant to corrosion. Vanadium strengthens steel whilst niobium is used in pacemakers. Dubnium (like lawrencium in Group 3 and rutherfordium in Group 4) is a member of the unstable transuranics (elements with an atomic number greater than 92; *see* Actinide and Transactinide Series, pages 152–3).

Group 6 – The Transition Metals: chromium (Cr), molybdenum (Mo) and tungsten (W)

These are refractory metals, being resistant to erosion and heat. The presence of chromium traces is what adds colour to rubies and emeralds. Chromium is frequently used in paint products, and also in the tanning industry. Although molybdenum had been used since ancient times to strengthen steel, by the 19th century it was deemed too brittle; then in 1906 the American physicist William D. Coolidge patented a method for rendering it ductile, since when it has returned to favour.

Group 7 – The Transition Metals: manganese (Mn), technetium (Tc), rhenium (Re) and bohrium (Bh)

The presence of two electrons in the outer shell of these elements ensures their similar chemical qualities. Technetium and rhenium both fill gaps left by Mendeleyev in his original table. The former was the first element to be discovered only after it had been created artificially (in a particle accelerator). Manganese is an essential trace element in the human body, although excess ingestion can prove toxic. Bohrium is so rare that it has never been isolated in its pure form, and consequently its properties have not yet been scientifically established (although, of course, they are predictable, thanks to Mendeleyev).

Group 8 - The Transition Metals: iron (Fe), ruthenium (Ru), osmium (Os) and hassium (Hs)

This group includes iron, which is one of the best-known elements on the planet. Iron was one of the earliest worked metals, after which the Iron Age is named (beginning around 1200 BC). Its red oxide also accounts for the appearance of the 'Red Planet' Mars.

Group 9 – The Transitional Metals: cobalt (Co), rhodium (Rh), iridium (Ir) and meitnerium (Mt)

The rare element, cobalt, was first discovered in the third millennium BC by the Ancient Egyptians and the Babylonians, who used its bluish hue to colour jewellery. Rhodium and iridium are the rarest naturally occurring elements on earth, with traces being found mixed with platinum. These were not discovered until the early 19th century.

Group 10 – The Transitional Metals: nickel (Ni), palladium (Pd) and platinum (Pt)

Medieval Germans named nickel after a demon said to inhabit mineshafts. It is now commonly used as an alloy to provide resistance to corrosion. Platinum – originally disregarded by the 15th-century Spanish explorers who came across it in South America – is now regarded as more valuable than gold.

Group 11 – The Transitional Metals : copper (Cu), silver (Ag), gold (Au) and roentgenium (Rg)

These are three of the best-known metals, ascending in weight, rarity and value. Copper is second only to silver as a conductor of electricity, hence its common use in wiring. Gold, silver and (to a lesser extent) copper have been used in coinage since ancient times. They are not intrinsically more valuable, but their gleaming appearance, relative scarcity and resistance to erosion have proved attractive qualities.

Group 12 – The Transitional Metals: zinc (Zn), cadmium (Cd), mercury (Hg) and coppernicum (Cn)

Often known as the volatile metals, this group includes zinc, which is an essential trace element in organic and animal life, including humans, who can absorb it from oysters. Mercury was highly valued by alchemists on account of its silvery, fluid properties, which were thought to indicate magical qualities.

GROUPS 13-18

The characteristics and members of these groups, from the boron group to the noble (or inert) gases are described in much greater detail in the main text.

Group 13 – The Boron Group: boron (B), aluminium (Al), gallium (Ga), indium (In) and thallium (Tl)

With the exception of boron itself, which is a semi-metal, all of the elements in this group are silvery white metals; all of the elements – including boron – have a valency of three.

Group 14 – The Carbon Group: carbon (C), silicon (Si), germanium (Ge), tin (Sn) and lead (Pb)

This group contains some of the most commonly occurring elements, including carbon, tin and lead (all used since ancient times). Among the usual periodic trends is increased metallic properties: carbon is a non-metal and silicon largely non-metallic, whereas lead and tin have wholly metal characteristics.

Group 15 – The Nitrogen Group: nitrogen (N), phosphorus (P), arsenic (As), antimony (Sb) and bismuth (Bi)

As with the Group 14, metallic properties increase from nitrogen (a colourless gas) through to bismuth (a metallic solid). Each has a valency of five, indicating the number of electrons in its outer shell.

Group 16 – The Oxygen Group: oxygen (O), sulphur (S), selenium (Se), tellurium (Te) and polonium (Po)

Also known as the chalcogens, whereby this group takes its name from the Greek word *chalcos* and the Latinized Greek word *genes* meaning 'copper-former' or, more generally, 'ore-former'. Under this nomenclature, oxygen itself is sometimes excluded from the group due to its very different behaviour from the other elements. All of the elements in this group have a valency of six.

Group 17 – The Halogens: fluorine (F), chlorine (Cl), bromine (Br), iodine (I) and astatine (At)

Sometimes known as the fluorine group, all of these elements form acids when bonded to hydrogen. Elements in the middle of the group are all commonly used as disinfectants. The halogen group is the only one in the periodic table to contain elements in all three common states (that is liquid, solid and gas). All of the halogens form acids when bonded to hydrogen.

Group 18 – The Noble Gases: helium (He), neon (Ne), argon (Ar), krypton (Kr), xenon (Xe) and radon (Rn)

In their natural state, the noble gases are all colourless, odourless, tasteless and non-flammable. All are extremely stable, and consequently tend not to form naturally occurring compounds. The term *Edelgas* (German for 'noble gas') was coined by Hugo

Erdmann in 1898, the same year that William Ramsay isolated neon, krypton and xenon (*see* page 55). The group takes its name from the fact that these chemical elements are all single-atom gases with very low reactivity; they are also colourless and odourless. The discovery of the noble gases helped to develop new theories of atomic structure, in particular the discovery that the electrons of an atom are contained in a shell around its nucelus.

The Lanthanide Series: lanthanum (La), cerium (Ce), praseodymium (Pr), neodymium (Nd), promethium (Pm), samarium (SM), europium (Eu), gadolinium (Gd), terbium (Tb), dysprosium (Dy), holmium (Ho), erbium (Er), thulium (Tm), ytterbium (Yb)

These are the metallic elements with atomic numbers ranging from 57 through to 70, and they are sometimes referred to as the 'rare earths'. The lighter elements in the series are chemically similar to the lightest, lanthanum, which gives the group its name. As with other groups, these elements (with the exception of europium) have ascending densities, from just over 6 grammes per cubic centimetre to just under 10 grammes per cubic centimetre. They also have a tendency to higher melting points as their atomic number ascends. Lanthanum compounds were originally used in carbon arc lamps. Some versions of the periodic table also include lutetium (Lu) with atomic number 71.

The Actinide Series – atomic numbers 89–103: actinium (Ac), thorium (Th), protactinium (Pa), uranium (U), neptunium (Np), plutonium (Pu), americium (Am), curium (Cm), berkelium (Bk), californium (Cf), einsteinium (Es), fermium (Fm), mendelevium (Md) and lawrencium (Lr)

These are all heavy metals and are radioactive, meaning their nuclei are liable to spontaneous disintegration. They are rare, as well as dangerous on account of their radioactivity. When the large nucleus of uranium (U) is bombarded with slow-moving neutrons, these are absorbed by the nucleus (instead of bouncing off it). The additional particles make the heavier nuclei unstable, causing them to split and release further, slow-moving neutrons. These further neutrons repeat the process with other neighbouring nuclei, setting off a self-sustaining nuclear reaction that can release immense amounts of explosive power (or, alternatively, be harnessed to produce a controlled source of nuclear energy).

The next element in the group, neptunium (Np), is the last naturally occurring element. Above this, other elements can only be created under extreme laboratory conditions. The transuranic elements, those with greater atomic number than 92, are unstable, having half-lifes ranging from fractions of a second to millions of years.

The Transactinide Series – atomic numbers 104–118; rutherfordium (Rf) to ununoctium (Uuo)

These are extremely rare elements with higher atomic numbers than the actinides. According to one convention, these begin with lawrencium (atomic number 103). Some of the highest transactinides, such as ununtrium (Latin for 113, its predicted atomic number; it is also known as eka-thallium) have been claimed but not verified or generally accepted. Only then will they be given a 'proper' name; those elements named to date (by ascending atomic numbers 104–110) are rutherfordium (Rf), dubnium (Db), seaborgium (Sg), bohrium (Bh), hassium (Hs), meitnerium (Mt) and darmstadtium (Ds).

A SIMPLIFIED PERIODIC TABLE

In some instances – and especially in schools – the periodic table is configured as groups 1–7 and 0, where the group number corresponds to the number of electrons in the outer shell of all the elements in that group.

1	2							
								H 1
Li 3	**Be** 4							
Na 11	**Mg** 12							
K 19	**Ca** 20	**Sc** 21	**Ti** 22	**V** 23	**Cr** 24	**Mn** 25	**Fe** 26	**C** 27
Rb 37	**Sr** 38	**Y** 39	**Zr** 40	**Nb** 41	**Mo** 42	**Tc** 43	**Ru** 44	**R** 45
Cs 55	**Ba** 56	**La** 57	**Hf** 72	**Ta** 73	**W** 74	**Re** 75	**Os** 76	**I** 77
Fr 87	**Ra** 88	**Ac** 89						

	3	4	5	6	7	0
						He 2
	B 5	**C** 6	**N** 7	**O** 8	**F** 9	**Ne** 10
	Al 13	**Si** 14	**P** 15	**S** 16	**Cl** 17	**Ar** 18
li **Cu** 29	**Zn** 30	**Ga** 31	**Ge** 32	**As** 33	**Se** 34	**Br** 35	**Kr** 36
d **Ag** 47	**Cd** 48	**In** 49	**Sn** 50	**Sb** 51	**Te** 52	**I** 53	**Xe** 54
't **Au** 79	**Hg** 80	**Tl** 81	**Pb** 82	**Bi** 83	**Po** 84	**At** 85	**Rn** 86

LIST OF RECOMMENDED FURTHER READING

- Philip Ball *The Ingredients: A Guided Tour of the Elements* (Oxford University Press, 2002)

A readable elaboration of the elements and their discovery.

- John Emsley *Nature's Building Blocks,* new rev edn (Oxford University Press, 2011)

A vast (720-page) compendium of the elements, listing their properties and a host of interesting facts about the discovery and use of all known elements. Addictively readable: ideal for dipping into – as one critic found, 'it's like a bar of chocolate, you can't eat just one square'.

- Sam Kean *The Disappearing Spoon of madness, love and the History of the World:* And Other True Tales from the Periodic Table (Black Swan, 2011)

An amusing run-through of the elements, along with a host of anecdotes and stories.

- Primo Levi *The Periodic Table* (London 1975)

A literary take on the subject: each of the 21, autobiographical short stories takes its name from and connects in some way to a chemical element.

- Linus Pauling *The Nature of the Chemical Bond* (Cornell University Press, 1960)

An introduction to modern structural chemistry by the finest chemist of the 20th century. An academic work, it rewards attention.

- Eric R Scerri *The Periodic Table: Story and its Significance* (Oxford University Press, 2007)

A perceptive, academic elaboration of the history of the periodic table.

- Paul Strathern *Mendeleyev's Dream: The Quest for the Elements* (Hamish Hamilton Ltd, 2000)

A prize-winning history of the discovery of the elements.

INDEX

absolute zero 109
acidic propensity 61, 64–5
actinide series 152
actinium (Ac) 152
air 8, 72, 78
alabamine (Ab) 62
alkali earth metals 41, 61, 144
alkali metals 29, 32, 61, 106, 144
Allison, Fred 62
allotropes 141
alpha particles 120–1
aluminium (Al) 36, 43, 149
americium (Am) 152
ammonia (NH3) 47
antimony (Sb) 78, 149
argon (Ar) 54–5, 104, 105, 124, 150
Arrhenius, Svante 119
arsenic (As) 78, 80, 82, 149
Asimov, Isaac 90
astatine (At) 60, 62, 138, 150
atomic density 37, 42
atomic mass 57
atomic numbers 33, 57, 60, 121, 124, 129
atomic weight 20, 28, 33, 37, 42–3, 51
atoms 70, 117, 120
  nuclear fusion 66
  radioactive decay 137
  splitting the 125
  structure 33, 129, 151
  valency 46–8

barium (Ba) 89, 95, 125, 144
Becquerel, Henri 116, 119
benzene 23, 47–8
berkelium (Bk) 152
beryllium (Be) 32, 41, 43, 60, 61, 125, 144
bismuth (Bi) 78, 82, 149
bohrium (Bh) 147, 153
Böhtlingk, Otto von 39
boron (B) 36, 43, 60, 149
boron group 36, 60, 149
Boyle, Robert 70–3, 75, 78, 79, 84, 95, 101
Brand, Hennig 79, 84, 88
bromine (Br) 28–9, 60, 107, 150

cadmium (Cd) 41, 148
caesium (Cs) 29, 32, 42
calcium (Ca) 144
californicum (Cf) 152
carbon (C) 17, 23, 40, 43, 47–8, 110, 141, 149
carbon dating 134–5, 141
carbon-14 134, 137, 141
carbon group 60, 149
Cavendish, Henry 92–3, 101, 123
Cavendish Laboratory, Cambridge 117, 123
cerium (Ce) 151
cesium (Cs) 144
chalcogens 150
chlorine (Cl) 28–9, 60, 124, 150

atomic weight 32, 33, 107
  discovery of 89, 90
chromium (Ch) 146
cobalt (Co) 147
compounds 71, 72, 98
Comte, Auguste 35
Coolidge, William D. 146
copernicium (Cn) 135
Copernicus, Nicolaus 135
copper (Cu) 78, 148
copper (coinage) group 60
Curie, Marie 116–17, 119, 138
Curie, Pierre 116–17, 119
curium (Cm) 138, 152

darmstadtium (Ds) 153
Davy, Humphry 90
deuterium (D) 94–5
deuterons 133
Döbereiner, Johann 106, 110–11, 113
Dorn, Friedrich Ernst 55
dubnium (Db) 127, 146, 153
dysprosium (Dy) 151

earth 8, 72, 78
Einstein, Albert 70, 123, 125
einsteinium (Es) 138, 152
electrons 101, 117, 120–1, 123, 124, 129, 138, 151
elements 70–3, 78, 98–9

Emsley, John 138–9
energy, nuclear 66, 125, 152
erbium (Er) 151
Erdman, Hugo 150
europium (Eu) 151

Fermi, Enrico 125, 127, 132, 133
fermium (Fm) 127, 152
Feynman, Richard 25, 138
fire 8, 72, 78
flerovium (Fl) 149
fluorine (F) 28–9, 43, 60, 119, 124, 150
fluorine group 150
francium (Fr) 144

gadolimium (Gd) 151
gallium (Ga) 37, 40, 149
Geiger, Hans 120
germanium (Ge) 40–1, 149
Godfrey, Ambrose 84
Goethe 113
gold (Au) 17, 72, 78, 79, 98, 148
Greeks, Ancient 71, 72, 78, 117

hafnium (Ha) 145
Hahn, Otto 124–5, 132
half-lives 134–5, 137, 152
halogen group 28–9, 64, 124, 150
astatine (At) 60, 62, 150
atomic weights 32, 107
hassium (Hs) 153

helium (He) 57, 60, 104, 124, 150
discovery of 55
nuclear fusion 66
properties 106
holmium (Ho) 151
hydrochloric acid (HCl) 64–5, 98, 107, 110
hydrogen (H) 99, 101, 105, 107, 144
atomic weight 28, 32, 101
benzene ring 23, 47–8
isotopes 94–5
nuclear fusion 66
position within the periodic table 51, 57, 60, 61
water 20, 46, 93
hydrogen group 144
hydroxides 107

indium (In) 149
inert gases 105, 121, 124
inorganic chemistry 48–9
iodine (I) 28–9, 107, 150
iridium (Ir) 42, 147
iron (Fe) 17, 78, 147
isotopes 94–5

Joliot-Curie, Irène 119, 127, 132
joliotium 127

Kekulé, August 23, 47–8
Klaproth, Martin 94–5, 145
Klason, Peter 119
Kreuger, Ivar 85

krypton (Kr) 55, 95, 104, 150–1

lanthanide series 151
lanthanum (La) 125, 151
Lavoisier, Antoine 8, 93, 98–9, 101
law of octaves (Newlands) 15
Lawrence, Ernest 145
lawrencium (Lr) 145, 146, 152, 153
lead (Pb) 17, 78, 110, 149
Lecoq de Boisbaudran, Paul-Émile 37, 40
Linnaeus, Carl 98
lithium (Li) 29, 32, 42, 144
position within the periodic table 51, 57, 60, 61, 106–7
livermorium (Lv) 150
lutetium (Lu) 145, 151

magnesium (Mg) 41, 54, 144
manganese (Mn) 89, 132, 146–7
Maxwell, James Clerk 123
Meitner, Lise 124–5, 135
meitnerium (Mt) 135, 137, 153
mendelevium (Md) 25, 138, 152
Mendeleyev, Dmitri Ivanovich:
discrepancies in periodic table 33, 36–7, 40–1

early life 8–9, 12–13
early periodic table 11, 15–28, 31, 50–1, 63, 70, 104, 111, 116
eight principles 42–3
missing elements 145, 147
Nobel Prize 119
mercury (Hg) 78, 110, 148
Meyer, Julius Lothar von 15
Moissan, Henri 119
molybdenum (Mo) 89, 133, 146

neodymium (Nd) 151
neon (Ne) 55, 60, 104, 124, 150–1
neptunium (Np) 152
neutrons 33, 57, 94, 129
Newlands, John Alexander Reina 15
Newton, Isaac 8, 73, 75, 95
nickel (Ni) 148
niobium (Nb) 133, 146
nitrogen (N) 43, 48, 54, 80, 82, 104, 110, 149
nitrogen group 60, 80, 82, 149
noble gases 121, 124, 129, 138, 150–1
discovery of 54–5, 104–5
position within periodic table 60
nuclear fission 66, 95, 125
nuclear fusion 66, 96

nucleus 33, 57, 66, 94, 121, 129, 151

organic chemistry 48–9
osmium (Os) 42, 147
oxygen (O) 43, 105, 107, 110, 150
discovery of 89, 90
valency 46, 48
water 20, 46, 93
oxygen group 60, 150

palladium (Pd) 148
phlogiston 89, 92
phosphates 82–3
phosphorous (P) 79–85, 88, 89, 99, 149
platinum (Pt) 42, 92, 138, 148
plutonium (Pu) 152
polonium (Po) 117, 119, 150
potassium (K) 29, 42, 106–7, 144
praseodymium (Pr) 151
Priestley, Joseph 90
promethium (Pm) 151
protactinium (Pa) 152
protons 33, 57, 94, 101, 129

quantum mechanics 121, 123

radioactive decay 137
radioactivity 116–17, 119, 120, 124–5, 127, 152
radiometric dating 134–5

radium (Ra) 117, 119, 144
radon (Rn) 55, 60, 104, 150
Ramsay, William 13, 54–5, 104, 150–1
Rayleigh, Lord (John William Strutt) 54–5
rhenium (Re) 132–3, 146–7
rhodium (Rh) 147
Roentgen, Wilhelm 116
roentgenium (Rg) 138
Royal Society of London 75, 93
rubidium (Rb) 29, 42, 144
Russian Chemical Society 19, 28, 42
Rutherford, Ernest 120–1, 123
rutherfordium (Rf) 138, 145, 146, 153
ruthernium (Ru) 147

salt (NaCl) 65, 107
samarium (Sm) 151
scandium (Sc) 145
Scerri, Eric 111
Scheele, Carl 85, 88–90
seaborgium (Sg) 153
Segrè, Emilio 133
selenium (Se) 41, 150
silicon (Si) 40, 43, 149
silver (Ag) 78, 110, 148
Smithsonian Institute 54–5
sodium (Na) 29, 61, 106–7, 144

sodium chloride (NaCl) 29, 32
sodium hydroxide (NaOH) 61, 64, 65, 107
spectroscopy 35, 37
strontium (Sr) 144
subatomic particles 129
sulphur (S) 78, 98, 150
supernova 96

tantalum (Ta) 146
technetium (Tc) 133, 146–7
tellurium (Te) 150
terbium (Tb) 151
thallium (Tl) 29, 149
Thomson, Sir Joseph John (J J) 117, 120, 123

thorium (Th) 152
thulium (Tm) 151
tin (Sn) 78, 149
titanium (Ti) 145
transactinide series 153
transition metals 132, 145–8
transuranic elements 127, 135, 146
triadic theory 106–7, 110–11
tritium 94–5
tungsten (W) 146

unbiseptium (Ubs) 135
ununoctium (Uuo) 153
ununtrium 153
uranium (U) 94–5, 99, 117, 124–5, 127, 152

valence shell 80
valency 46–8, 64
vanadium (V) 146

water (H2O) 8, 61, 64, 65, 72, 78
  elements in 20, 46, 93
Winkler, Clemens 40–1
Wöhler, Friedrich 49

x-ray 116
xenon (Xe) 55, 104, 150

ytterbium (Yb) 151
yttrium (Y) 145

zinc (Zn) 41, 148
zinc group 60
zirconium (Zr) 145

ACKNOWLEDGEMENTS

I would particularly like to thank my editor Simon Willis for all his painstaking work, suggestions and so forth. Well beyond the call of duty. Also Jane O'Shea and all the team at Quadrille, who have contributed so much.

	1	2			3	4	5	6	7	
PERIODS										
1	HYDROGEN 1 **H** 1.0079									
2	LITHIUM 3 **Li** 6.941	BERYLLIUM 4 **Be** 9.0122								
3	SODIUM 11 **Na** 22.990	MAGNESIUM 12 **Mg** 24.305								
4	POTASSIUM 19 **K** 39.098	CALCIUM 20 **Ca** 40.078		SCANDIUM 21 **Sc** 44.956	TITATIUM 22 **Ti** 47.867	VANADIUM ·23 **V** 50.942	CHROMIUM 24 **Cr** 51.996	MANGANESE 25 **Mn** 54.938	IR 2 **F** 55
5	RUBIDIUM 37 **Rb** 85.468	STRONTIUM 38 **Sr** 87.62		YTTRIUM 39 **Y** 88.906	ZIRCONIUM 40 **Zr** 91.224	NIOBIUM 41 **Nb** 92.906	MOLYBDENUM 42 **Mo** 95.94	TECHNETIUM 43 **Tc** [98]	RUTH 4 **R** 10
6	CAESIUM 55 **Cs** 132.91	BARIUM 56 **Ba** 137.33	57-70 *	LUTETIUM 71 **Lu** 174.97	HAFNIUM 72 **Hf** 178.49	TANTALUM 73 **Ta** 180.95	TUNGSTEN 74 **W** 183.84	RHENIUM 75 **Re** 186.21	OS 7 **C** 19
7	FRANCIUM 87 **Fr** [223]	RADIUM 88 **Ra** [226]	89-102 **	LAWRENCIUM 103 **Lr** [262]	RUTHERFORDIUM 104 **Rf** [261]	DUBNIUM 105 **Db** [262]	SEABORGIUM 106 **Sg** [266]	BOHRIUM 107 **Bh** [264]	HAS 1 **H** [

★ LANTHANIDE SERIES

LANTHANUM 57 **La** 138.91	CERIUM 58 **Ce** 140.12	PRASECDYMIUM 59 **Pr** 140.91	NEODYMIUM 60 **Nd** 144.24	PROMETHIUM 61 **Pm** [145]	SAM **S** 1

★★ ACTINIDE SERIES

ACTINIUM 89 **Ac** [227]	THORIUM 90 **Th** 232.04	PROTACTINIUM 91 **Pa** 231.04	URANIUM 92 **U** 238.03	NEPTUNIUM 93 **Np** [237]	PLU **F**